IV Russian Forum for Young Scientists with International Participation “Space Engineering”

MATEC Web of Conferences Volume 48 (2016)

Tomsk, Russia
12 - 14 April 2016

Editors:

Valeriy N. Borikov
Antonia G. Dolgih
Veronika S. Ivanova
Anastasiya A. Ignatovskaya

ISBN: 978-1-5108-2213-9

Printed from e-media with permission by:

Curran Associates, Inc.
57 Morehouse Lane
Red Hook, NY 12571

Some format issues inherent in the e-media version may also appear in this print version.

Printed by Curran Associates, Inc. (2016)

For additional information, please contact EDP Sciences – Web of Conferences
at the address below.

EDP Sciences – Web of Conferences
17, Avenue du Hoggar
Parc d'Activité de Courtabœuf
BP 112
F-91944 Les Ulis Cedex A
France

Phone: +33 (0) 1 69 18 75 75
Fax: +33 (0) 1 69 28 84 91

contact-edps@webofconferences.org

Additional copies of this publication are available from:

Curran Associates, Inc.
57 Morehouse Lane
Red Hook, NY 12571 USA
Phone: 845-758-0400
Fax: 845-758-2633
Email: curran@proceedings.com
Web: www.proceedings.com

TABLE OF CONTENTS

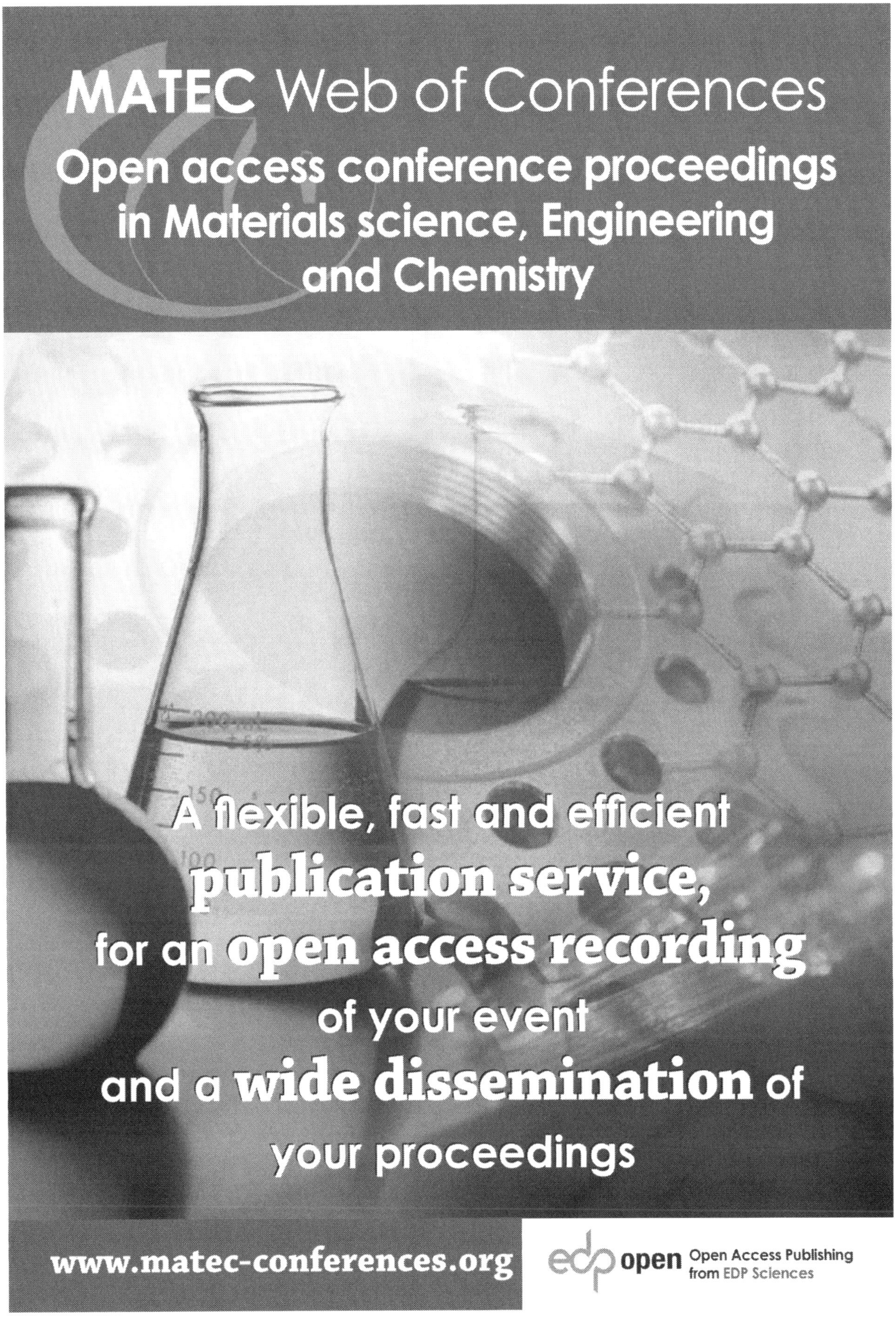
MATEC Web of Conferences
Open access conference proceedings in Materials science, Engineering and Chemistry
A flexible, fast and efficient publication service, for an open access recording of your event and a wide dissemination of your proceedings
www.matec-conferences.org
edp open Open Access Publishing from EDP Sciences

ISSN (electronic edition): 2261-236X

edp sciences

MATEC Web of Conferences

Statement of Peer review

In submitting conference proceedings to MATEC Web of Conferences, I certify to the Publisher that I adhere to the **Policy on Publishing Integrity** of the journal in order to safeguard good scientific practice in publishing.

1. All articles have been subjected to peer review administered by the proceedings editors.
2. Reviews have been conducted by expert referees, who have been requested to provide unbiased and constructive comments aimed, whenever possible, at improving the work.
3. Proceedings editors have taken all reasonable steps to ensure the quality of the materials they publish and their decision to accept or reject a paper for publication has been based only on the merits of the work and the relevance to the journal.

Title, date and place of the conference

IV Russian Forum for Young Scientists with International Participation „Space Engineering"
12-14 April 2016
Tomsk, Russia

Proceedings editor(s):

Valeriy N Borikov
Antonina G. Dolgih
Veronika S. Ivanova
Anastasiya A Ignatovskaya

Date and editor's signature

04.03.2016

MATEC Web of Conferences 48, 01001 (2016)
DOI: 10.1051/matecconf/20164801001

Design and analysis of power supply and information transfer to three-axis gyroscope stabilizer platform

Lev Belyanin [1,a], Doan Ket Vu [1], Timofey Pozharskiy[1] and Trong Yen Nguyen[2]

[1] *National Research Tomsk Polytechnic University, 634050 Tomsk, Russia*
[2] *Science and Technology Institute of Military, No. 17 Hoang Sam, Nghia Do, Cau Giay District, Hanoi, Viet Nam*

Abstract. The paper focuses on problems of the engineering implementation of 3-axis gyrostabilizer of the camera for orthophotography performed from light and ultralight aircrafts. The functional diagram of gyrostabilizer is characterized by a lighter ball-bearing having a large inner diameter used for the platform suspension on its internal (vertical) axis. The problem is to transfer the supply voltage to the platform and the electric signals to and from the platform. Design solutions concerning the power supply and information transfer are supported by the experimental techniques suggested in this paper, namely: the use of the rolling contact; three-phase gyro motor power supply replaced by that from the single-phase network; signal transmission from the precession angle sensor using the air transformer; current and voltage frequency division; DC mode selection for the torquer. The results obtained prove the efficiency of the suggested design solutions.

1 Introduction

A three-axis gyroscopic stabilizer is intended for the digital camera stabilization to produce orthophotography from light and ultralight aircrafts, such as helicopters, airplanes and trikes. A vertical stabilization of the camera is necessary to generate a true vertical photograph taken with its optical axis such that an aircraft's position coordinates detected by the GPS or GLONASS receivers would correspond to the centre of each aerial photograph. An azimuth-stabilized camera, i.e. its vertical stabilization, firstly, improves the quality of aerial photographs and, secondly, creates the conditions for generating a continuous picture of the Earth surface. This gyroscopic stabilizer incorporated on aircraft can be simultaneously used as its course detector. The latter is important especially when a regular course device is not available on board.

Digital cameras are currently installed on gyroscopic stable reference platforms to produce orthophotography from light and ultralight and, hence, inexpensive aircrafts [1, 2].

Orthophotography is used in carrying out cadastral works, pipeline, highway and railway examination, and others.

[a] Corresponding author : belyaninln@tpu.ru

2 Schematic of gyroscopic stabilizer

The suggested gyrostabilizer is based on the principle of the powered gyroscopic stabilization using three restrained gyros having a ball-bearing suspension. The functional diagram of the gyrostabilizer is given in Figure 1 that does not include the camera.

The gyrostabilizer is characterized by the series stabilization circuit that stabilizes the internal frame of gimbal suspension using the two-gyro leveling system. A uniaxial stabilizer with the normal stabilized axis is placed on the internal frame. The platform suspensions are represented by the lighter ball-bearing having a large inner diameter.

In terms of the problem solution, this schematic is considered to be optimum due to the following reasons:

- to provide stabilization, only one restrained gyro should be placed on the platform. Thus, the residual space allows installing the camera;
- the camera lens is placed in the opening in the centre of the platform and available to an operator;
- coordinate resolver is not required;
- a Y-shaped cantilever used in the capacity of the external frame allows reducing the weight and dimensions of stabilizer and clear a space from parts and units in front of the camera lens.

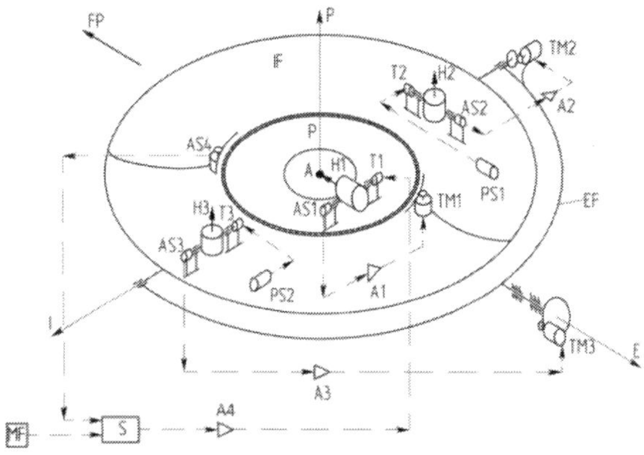

P – platform; *IF* – internal frame of gimbal suspension; *EF* – external frame of gimbal suspension; *G* – gyros; *AS* – angular direction sensors; *T* – torquers; – vectors of kinetic moment; *PS* – pendulum sensors; *A* – amplifiers; *TM* – torque motors; *MF* – magnetic field sensor; *S* – summer; *AP, AI, AE* – suspension axes of platform, internal and external frames, respectively; *FP* – flight path.

Figure 1. Functional diagram of gyrostabilizer.

In the stabilization channel of azimuth-stabilized platform, the gyrostabilizer is corrected by the magnetic field sensor placed beyond the gyrostabilizer.

The gyrostabilizer unit includes: gyrostabilizer proper; electronic assembly; magnetic field sensor in azimuth correction system.

The electronic assembly comprises the secondary power supply; power supply converter of gyros, angular direction sensors and torquers; torque motor amplifiers with correction circuits; azimuth correction system of the platform.

3 Problem definition

The gyroscopic syabilizer is desiged to operate at roll and pitch angles not exceeding 30 degrees. Therefore, flexible leads placed inside the hollow semi-axes of suspension are used on its axes of the external and internal frames. The angle of the platfrom rotation is not restricted relative to the internal frame, hence the flexible leads cannot be used on the platform suspension axis. In this case, sliding current collectors are conventionally used, which is a combination of slip rings placed on the axis of rotation and current collectors adjusted to them [3]. With a view to reduce the friction torque arising in this sliding current collector and increase its wear resistance, the diameter of slip rings is selected as less as possible (in real gyroscopic instruments, the diameter of slip rings comes to millimeter units in the best models of current collectors).

Even a shallow analysis of the schematic presented in Figure 1, shows the impossibility of using sliding current collectors on the suspension axis of the platform due to the large diameter of slip rings and their large number. In case of the traditional materials and values of contact pressures used in the construction of the sliding current collectors, the slip ring diameter increased up to 130-150 mm enables the increase of the friction torque almost by an order. The friction torque in suspensions, sliding current collector, and torque motor gear is the main disturbing factor that reduces the accuracy of the azimuth-stabilized platform. Hence, the success in the whole project depends on the successful design solution of sliding current collectors.

This paper presents results of the development and investigation of the platform power supply, signal transmission from the platform to the gyrostabilizer base and from the internal frame to the platform.

4 Main design solutions

According to Figure 1, the restrained gyro is placed on the platform by the suspension axis of which the precession angle sensor and torquers are installed. Let us give the important characteristics of the gyro motor and sensors.

The asynchronous gyro motor operates on a 36V alternating current (AC) 400 Hz three-phase power supply.

The induction precession angle sensor has the exciting winding which operates on a 36VAC 400 Hz power supply.

The AC electromagnetic torquer has the exciting windings which also operate on a 36VAC 400 Hz power supply. AC voltage 400 Hz supplied to the control winding is phased with the voltage supplied to the exciting winding.

Since the exciting windings are supplied by voltages similar to that of the gyro motor, seven circuits of the current transfer are required to provide the functioning of the platform azimuth-stabilizing and correcting channels.

Therefore, the authors suggest the following.

First, to provide only two circuits for the electric coupling between the platform and the internal frame. To implement one of them (case circuit), a ball-bearing of the platform suspension is suggested which can successfully perform this function due to the large number of balls. The second circuit is based on a rolling contact, i.e. an assembly comprising a ring with a large diameter and a roller adjusted to it. This roller is bound to be rotated in its supports with the minimum friction and has a proper electrical connection with its holder. These circuits provide the simultaneous of 36VAC 400 Hz single-phase and direct current (DC) power supply for the torquer control. It is advisable to use coupling capacitors and inductors to provide the efficient frequency division of AC and DC supply voltage.

Second, the gyro motor three-phase power supply is advisable to replace by an AC single-phase power supply using a phase-shifting capacitor. The experience of this application is described in works [4, 5, 6].

Third, to provide DC operation for the torquer to allow the exciting winding to operate on a DC rectifier. The latter is placed on the platform. Experiments show that the operation of this DC electromagnetic torquer is much better than that on AC supply.

Fourth, the wireless transmission system should be used to transmit the AC output signal from the angular direction sensor to the base of the torque motor amplifier and then to its input.

The description of the experimental facility designed and manufactured to prove the efficiency of suggestions and their experimental studies is given in Section 5.

5 Experiments

Figure 2 presents the electric schematic diagram of the experimental facility intended for the investigation of the rolling contact with the frequency division of supply voltage and control signals transmitted.

Low resistor R_a is series-connected to the control winding of the torquer and intended for the oscillographic measurements of the current in this winding, including the control current variable.

The experimental facility is based on a tilt stand the platform of which arranges the slip ring, restrained gyro with the angular direction sensor and the torquer, and cells arranged in accordance with the electric schematic diagram shown in Figure 2.

The roller is mounted in the arm-holder of the journal bearing. The slip ring, roller, and its holder are made of the type LS-59-1 brass. The diameters of the slip ring contact surface, the roller, and the pivot of the journal bearing are respectively 220, 10, and 2,5 mm. The roller is adjusted to the slip ring by the arm-holder also mounted to the journal bearing and a spring on the support. The spring force is screw-adjusted.

Allowing for the active power generation, transmission coil TC was replaced by its equivalent – a fixed resistor.

Experimental research is aimed at the detection of:
- the contact resistance of the sliding current collector at different rotation rates of the platform and different forces of adjusting the roller to the slip ring;
- the quality of the frequency division of AC and DC supply voltage of the torquer and optimization of the network parameters;
- currents and voltages at the representative points of the diagram.

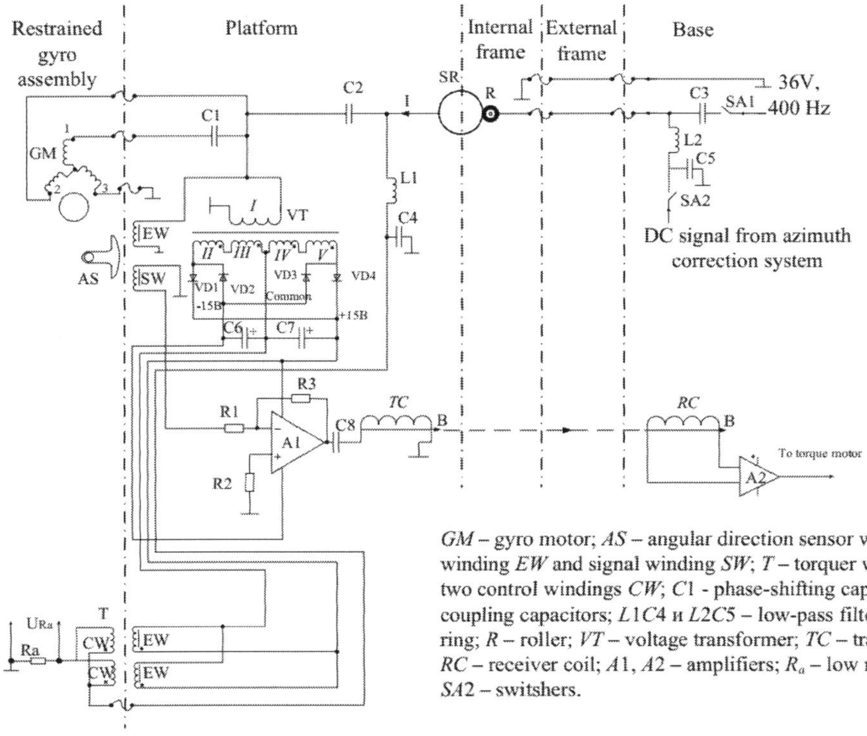

GM – gyro motor; *AS* – angular direction sensor with exciting winding *EW* and signal winding *SW*; *T* – torquer with two *EW* and two control windings *CW*; *C*1 - phase-shifting capacitor; *C*2, *C*3 – coupling capacitors; *L*1*C*4 и *L*2*C*5 – low-pass filters; *SR* – slip ring; *R* – roller; *VT* – voltage transformer; *TC* – transmission coil; *RC* – receiver coil; *A*1, *A*2 – amplifiers; R_a – low resistor; *SA*1, *SA*2 – switshers.

Figure 2. Electric schematic diagram of the experimental facility intended for the rolling contact investigation.

To detect the contact resistance R_c, the direct current is passed through the sliding current collector, while the loss of voltage U_c is registered by an oscillograph (Fig. 3). In case the current I is constant, the input voltage U will be proportional to the contact resistance R_c.

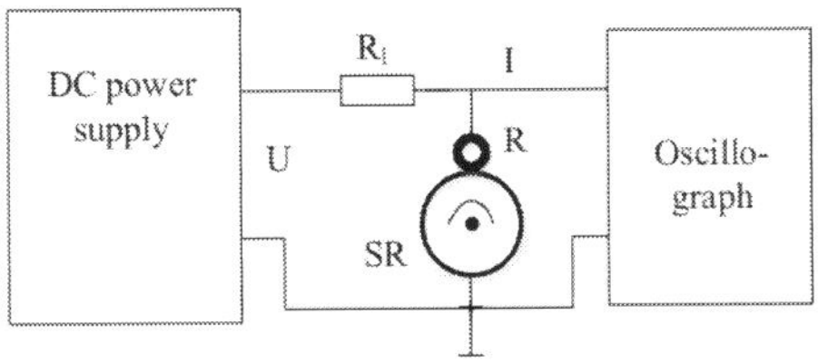

R_l – load resistor;
I – current;
U – output voltage;
SL – slip ring;
R – roller.

Figure 3. Schematic of contact resistance.

In order to provide the direct current passing through the sliding current collector, i.e. its supply from the current generator, the high output voltage U is used. The load resistor R_l is seriesly connected to the circuit. Voltage U_c is then obtained from

$$U_c = \frac{U}{R_l + R_c} \cdot R_c . \tag{1}$$

Since $R_l >> R_c$ then

$$R_c \approx \frac{R_1}{U} \cdot U_c . \tag{2}$$

The type GUA-2D gyro unit is used in the experimental facility. The following experimental parameters are used: 1 μF phase-shifting capacitance $C1$; 50 μF coupling capacitance $C2$; 44 μF coupling capacitance $C3$; $L1$=$L2$=5 H inductance of filter inductor; $C4$=$C5$=9,4 μF filter capacitance.

At voltage U = 62 V (Fig. 3) supplied through the load resistor of $R_1 = 280$ ohm, I = 0,22 A. The contact resistance is detected at different values of the adjustmnet force and different rates of the platform rotation. Figure 4 presents the oscillogram of the contact resistance that reflects its behavior in time during the platform rotation. For comparison, Figure 4 contains the curve of the static contact resistance R_{st}.

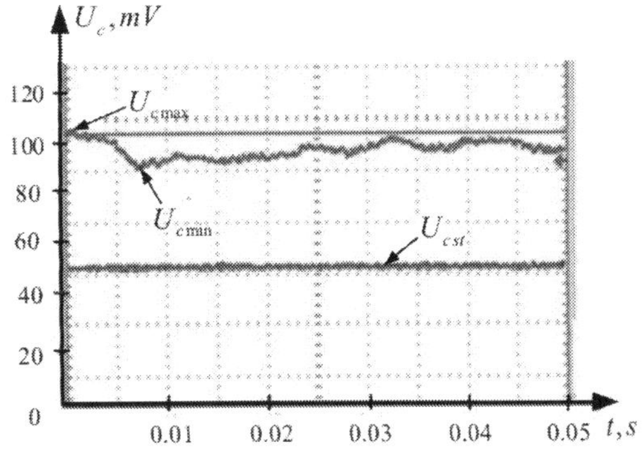

Figure 4. Contact resistance curves.

Research findings are given in Table 1 that includes the following notations: F – adjustment force; R_{st} – static contact resistance; ω – angular rate of rotation of the slip ring; R_{min} and R_{max} – minimum and maximum contact resistances within the range of which the slip ring is changed at its rotation.

According to Table 1, the slip ring rotation increases the contact resistance by 1.7 ÷ 2.6 times depending on the values of the adjustment force and the angular rate of rotation. The quality of contact is enhanced with the increase of the adjustment force and the decrease of the angular rate of rotation.

During the ring rotation, the contact resistance ranges between 0.05÷0.1 ohm that is quite acceptable in terms of the given problem solution.

Table 1. Experimental findings for contact resistance.

F, G_s	R_c, ohm	ω, dps	R_{min}, ohm	R_{max}, ohm	ΔR_c, ohm
100	0.385	20	0.873	0.946	0.073
		60	0.914	0.986	0.073
		120	0.914	0.986	0.073
150	0.287	20	0.568	0.623	0.055
		60	0.586	0.677	0.091
		120	0.718	0.827	0.109
200	0.280	20	0.464	0.536	0.073
		60	0.464	0.555	0.091
		120	0.518	0.609	0.091
250	0.250	20	0.446	0.518	0.073
		60	0.436	0.546	0.109
		120	0.505	0.596	0.091
300	0.254	20	0.427	0.482	0.055
		60	0.500	0.555	0.055
		120	0.500	0.573	0.073
350	0.236	20	0.418	0.473	0.055
		60	0.455	0.509	0.055
		120	0.486	0.541	0.054

Investigations of the power supply system of the gyro motor and other devices located on the platform shown in Figure 3, show the efficiency of design solutions implemented in the experimental facility.

Thus, the starting time of the gyro motor comes to 2 min and 55 s. The total consumption of the current passing through the sliding current collector is 0.22 A. Voltage losses on the coupling capacitors $C2$ and $C3$ are 1 and 1.3 V respectively. The current variable, i.e. an interference in the control winding of the torquer, is 0.247 mA at 10 mA, the maximum design value of the current correction.

6 Design and investigation of wireless transmission system

According to the functional diagram of gyrostabilizer, the signal transmission from the precession angle sensor placed on the platform is bound to be arrived to the input of the amplifier of the torque motor incorporated in the electronic assembly. The signal is transmitted from the platform to the base by means of the wireless transmission system, and further to the electronic assembly across the wire.

The following wireless transmission techniques are known: radio channel, optical channel, acoustic channel, variable magnetic field. The selection of the in-depth analysis technique and investigations was conditioned by the fact that the transmitted signal is an amplitude-modulated AC signal at 400 Hz carrier frequency.

The main requirements for the wireless transmission are as follows:
- stability of the transmission factor;
- no signal distortion;
- noiseimmunity;
- easiness of implementation in particular conditions.

The latter teqchnique was selected for further investigations. It is based on the use of the mutual induction of the transmission and receiver coils situated next to each other [7, 8]. The transmission coil placed on the gyrostabilized platform, accepts the enhanced signal from the precession angle sensor. The receiver coil placed on the gyrostabilizer base, generates the electromotive force that can be obtained from [9]:

$$e_2 = -\frac{d}{dt}(L_2 \cdot i_2 \pm M \cdot i_1), \qquad (3)$$

where e_2 is the electromotive force in the receiver coil; L_2 is the induction in the receiver coil; M is the mutual induction of coils; i_1 and i_2 are currents passing through coils.

In (3), the second component is the electromotive force of transformation, while the second is the electromotive force of self-induction. In case of a rather high input resistance of the amlifier, the first component is neglected.

Thus, the properties of the wireless transmission system mostly depend on the value and stability of the mutual induction of coils.

The induction depends on the geometry and the number of turns in both transmission and receiver coils, their mutual arrangement, and the environmetal conditions under which the variable magnetic field propagates, i.e. magnetic permeability and electrical conductivity.

The stability of the transmission factor in the wireless transmission system determines the stability of the transmission factor in the open circuit of stabilization loop in azimuth that ensures the accuracy of the system stabilization and consistency.

Theoretical studies and quantification of the different factors affecting the value of the mutual induction turned to be inefficient. Therefore, much experimetal research has been carried out into the following issues:
- to arrange the transmission and receiver coils in the gyrostabilizer such that to provide the minimum dependence between the mutual induction and the spatial position of the platform and gyrostabilizer base;
- to study the effect on the system operation exerted by metal parts and assemblies of gyrostabilizer that are situated nearby the transmission and receiver coils;
- to determine the value of the output resistance of the system as a source of signal.

To solve the first problem, a transmission coil and two variants of receiver coils were designed and manufactured.

The transmission coil was designed allowing for its mount position, i.e. on the platform under the ball-bearing suspension, close to it. The physical configuration of this coil is given in Figure 5.

The first variant of the receiver coil is a combination of four coils with series and cumulative connection placed on the base on the edge of the opening that gives the camera a field of view. The idea is to properly arrange the coils, thereby lowering the dependence between the total electromotive force of coils and the rotation of the transmission coil relative to the base. It is supposed that while the mutual induction of a receiver and a transmision coil increases at rotation, the induction of another coil placed on the other side of the opening decreases and *vice versa*. At the same time, this experiment shows the optimum coil arrangement relative to the base and the radius of the opening.

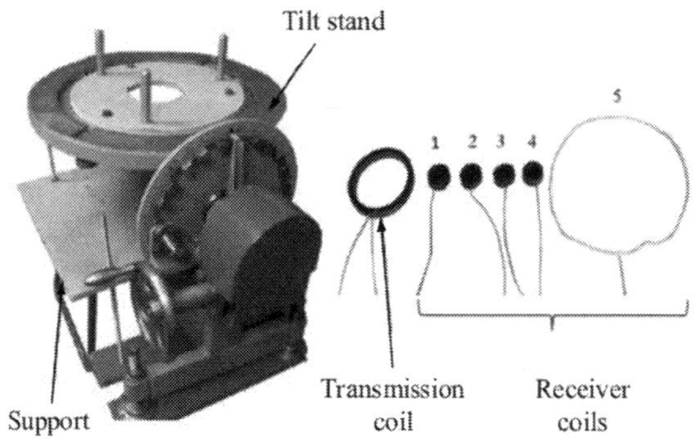

Figure 5. General view of the tilt stand and a set of coils.

The second variant of the receiver coil implies its increased diameter such that the total magnetic flux created by the transmission coil and tying the turns of the receiver coil, would less depend on the relative angular oscillations of coils. In Figure 5, it is the self-supported coil 5. The coil parameters are given in Table 2.

Table 2. Parameters of coils.

Coil types	Winding wire diameter, mm	Number of turns	Inner diameter of coil, mm	Resistance, ohm	Induction, mH
Transmission coil	0,4	172	109,3	9,1	6,82
Coil 1	0,1	650	30,3	155	19,1
Coil 2	0,1	650	30,3	156	19,1
Coil 3	0,1	650	30,3	159	19,7
Coil 4	0,1	650	30,3	155	19,1
Coil 5	0,14	150	310	216	34,1

Experiments on the optimization of the wireless transmission system are carried out using the uniaxial tilt stand shown in Figure 5. The transmission coil is mounted to the disk made of non-magnetic and non-conductive material. The disk, in turn, is mounted to the platform of the tilt stand. Receiver coils arranged on the support are also made of non-magnetic and non-conductive material. The distance between the transmission coil and the support is 240 mm that is equal to the distance between the coil and the gyrostabilizer base. Harmonic voltage 400 Hz is supplied to the transmission coil. The voltage of 3V is selected to provide 1W power generated by the coil. This voltage is constant. Receiver coils transmit voltage to the oscillograph and the AC electronic millivoltmeter. In each experiment, the transmission factor F is obtained as the ratio between the real value of voltage U_2 of receiver coils and that of the transmission coil:

$$F = \frac{U_2}{U_1}. \tag{4}$$

The transmission factor variation is presented in Figure 6 and has the following conditions. Four coils 1–4 shown in Figure 5, have a series and comulitive connection. The coil planes have different angles α of inclination relative to the support. The coils are mounted to the support in the centres of the four sides of the square. The square side equals to the diameter of the opening in the centre of gyrostabilizer. At a change in the angle of inclination of coils, their axes stay in the plane normal to the support that contains the opening axis.

The maximum value of transmission factor and its maximum independence of the angle of inclination β of the transmission coil are acheved at $\alpha = 90^0$, i.e. at the normal position of planes of coils 1-4 relative to the support. Then, the maximum change in the transmission factor is 4,2% at β angle of inclination ranging within $\pm 30^0$.

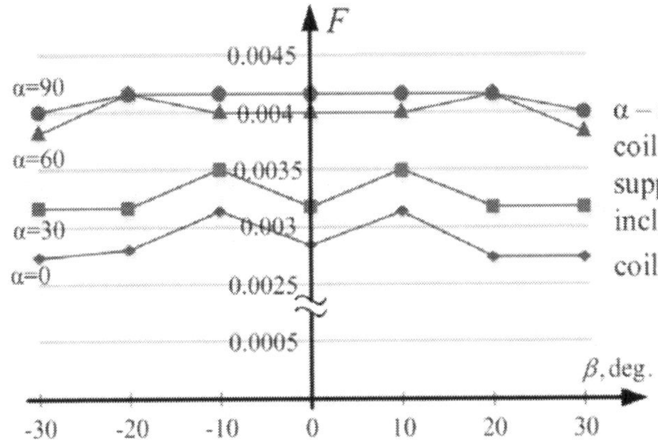

Figure 6. Transmission factor curves.

Experiments conducted for the self-supported coil 5 coaxially mounted to the support opening show the great stability of the transmission factor as well as its greater value. Its maximum changes come to 3,4% at β angle of inclination within the same range.

Since the second variant of the system with the self-supported coil provides the stability of transmission factor and is easier in its implementation, it is used in further experiments.

The dependence between the receiver coil voltage and the load current is obtained allowing to detect the internal resistance of the wireless transmission system as 222 ohm. In this experiment, the power supply has a low internal resistance. In case the transmission coil is supplied by the amplifier of the angular direction sensor, the internal resistance rather increases.

The experimental studies of the wireless transmission system are accomplished by the estimation of its behavior affected by metal parts and assemblies arranged nearby the transmission and receiver coils and between them.

As is known [10], the interaction between the variable magnetic field and metal is conditioned by the field frequency and metal properties, such as magnetic permeability and electrical conductivity. Magnetic permeability modifies the nature of the field/metal interaction at zero and low frequency. Magnetic field permeates through metal and becomes stronger in it. In this case metal acts as a magnetic circuit. At medium and high frequencies the nature of this interaction changes, and electrical conductivity becomes increasingly important. The variable magnetic field enables the electromotive force in metal mass resulting in eddy currents. The latter create their own variable magnetic field that interacts with the external field, therby weakening it in metal mass and strengthening beyond it.

In other words, the variable magnetic field created by the transmission coil is affected by any metal parts and assemblies that can lead to the change of transmission factor of the wireless system or unwanted phase shift between the voltage supplied to the transmission coil and that released by the receiver coil.

A ball-bearing with a large diameter is placed on the transmission coil side opposite to the receiver coil, that is used for the platform suspension. The transmission factor then decreases by 12%.

Two steel sheets of $\approx 0.5\,\text{kg}$ mass each are placed on the support nearby the reciever coil. The transmission factor then decreases by almost the same percentage.

The steel sheet 2 mm thick and area exceeding that covered by the receiver coil is placed between the coils. The wireless transmission system operates properly, however, the transmission factor decreases then by 37%.

As for non-magnetic metals such as aluminum and copper alloys placed nearby the coils, they have a small effect on the wireless transmission system.

These experimental findings prove the efficiency of the suggested wireless transmission system for the signal generated by the angular direction sensor. Probably, this is owing to the frequency value of signaling, i.e. 400 Hz. This frequency allows the air transformer formed by the two coils to consistently operate. However, the interaction between the variable magnetic field and conductive materials is still insufficient.

7 Conclusion

Research results proved the efficiency of the suggested design solutions that allows giving recommendations on their application to the gyrostabilizer and in other similar designs.

Experiments on the determination of the friction torque generated by the rolling contact have not been conducted. Taking into account the fact that the friction torque was mainly produced by the journal bearings of the roller suspension, the roller was mounted to its holder using ball-bearings, while the electric coupling between them was achieved by central point contacts. At a time, with a view to reduce the contact resistance and increase the contact reliability, two similar rollers were installed completely opposite to the axis of the platform suspension.

References

1. L. Cao, L. Hongbo, G. Kemper, ISPRS Annals **37**, 1667 (2008)
2. G. Kemper, ISPRS Annals **39**, 351 (2012)
3. E. Nikitin, S. Shestov, V. Matveev, *Gyro systems: gyroscope elements* (Vysshaya Shkola Publ., Moscow, 1988)
4. Y. Nguyen, K. Vu, *Proc. 18th Int. Sci. Conf. 'Modern Engineering and Technology* **1**, 381 (2012)
5. L. Belyanin, L. Gurin, V. Martemyanov, S. Samoylov, RF Patent 2003045 (1993)
6. L. Belyanin, A. Golikov, V. Martemyanov, S. Samoylov, *Coll. Papers OAO 'Geofit'* **2**, 34 (2002)
7. S. Cheng, P. David, *Sel. Papers Int. 10th Worksh. Micro and Nanotech. Power Gen. and Energy Convers. Appl.* ***61*** (2010)
8. J. Hirai, T. Kim, A. Kawamura, IEEE Trans. on Indust. Electronics **46**, 349 (1999)
9. L. Bessonov, *Teoreticheskie osnovy elektrotekhniki.* Elektricheskie tsepi [*Electrical Engineering Theory. Electric circuits*] (Vysshaya Shkola Publ., Moscow, 1996)
10. A. Nenashev, *Konstruirovanie radioelektronnykh sredstv* [*Radioelectronic equipment design*] (Vysshaya Shkola Publ., Moscow, 1996)

MATEC Web of Conferences 48, 01002 (2016)
DOI: 10.1051/matecconf/20164801002

Hybrid gravitational orientation system of small spacecraft

Victor Dmitriev[1], Ruslan Frolov[1,a] and Anastasia Bekasova[1]

[1] *National Research Tomsk Polytechnic University, 634050 Tomsk, Russia*

Abstract. In this article is considered the problem of small satellite's angular position control. This problem is solved by some improvement. The authors offer to use active control element, in the form of gyrodyne, which to fix on the end of gravitational rod. This improvement transforms the gravitational system of orientation into a control system along three axes of orientation: yaw, roll and pitch.

1 Introduction

The principle of gravitational stabilization of an object caused by the gradient of the gravitational field of the Earth has been known since the Lagrange's work about libration of the Moon published in 1780. He determined the conditions of stable oscillations of the body in vertical orientation of its longitudinal axis.

Thus, stabilization of the artificial earth satellite relative to the local vertical line can be realized using the properties of the Earth's gravitational field gradient with simultaneous damping of spacecraft librations without active technical control systems employed.

Under certain conditions (similar to the constant direction of the Moon relative to the Earth), a specified ratio of the moments of inertia of the spacecraft body will make the axis of smallest moment of inertia be constantly directed towards the center of the Earth.

The main advantage of the above orientation is that the aerospace directional antenna used for amplification of the spacecraft radiofrequency signal will by constantly directed to the center of the Earth.

Due to problems directly related to the development of space technology and the rate of technological evolution, the fundamental works on gravitational systems were published after 2000 [1-4].

2 Problem of modern gravitational orientation systems

The gravitational systems of orientation consist of a gravitational rod with load attached at its end. The shape of this small spacecraft can be considered as a dumbbell shape (the rod with a spacecraft and load fixed at its ends). During orbital flight, this spacecraft becomes the axis of the minimum moment of inertia perpendicular to the center of the Earth (figure 1). However, this method provides stabilization of the spacecraft only along 2 axes – pitch and roll. Along the yaw axis, the spacecraft remains free to spontaneously rotate about this axis at any angles. In some cases, this limits the

[a] Corresponding author : phrolov1994@mail.ru

performance of the spacecraft, for example, orientation of solar panels along the normal to the direction of sunlight.

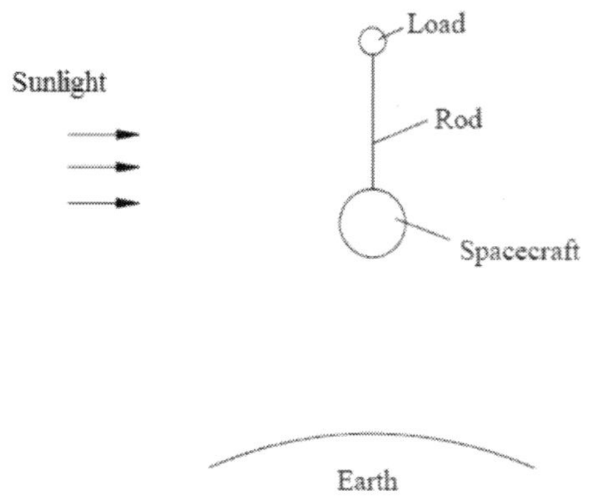

Figure 1. Principle of gravitational stabilization.

The main fundamental disadvantage of all gravitational systems of orientation is that passive gravitational systems of orientation are stabilizing systems of the angular position of the spacecraft, but not controlling ones.

3 Technical decision of the problem

This problem can be solved with different types of combined orientation systems [5]. This paper discusses a hybrid gravitational system of orientation of small spacecraft, where a two-axis gyroscope (gyrodyne) implements the function of the load. Its kinematic momentum is directed along the normal to the orbit plane, and the axis of the suspension is placed in the orbit plane, as shown in figure 2. This improvement transforms the gravitational system of orientation into a control system along three axes of orientation: yaw, roll and pitch.

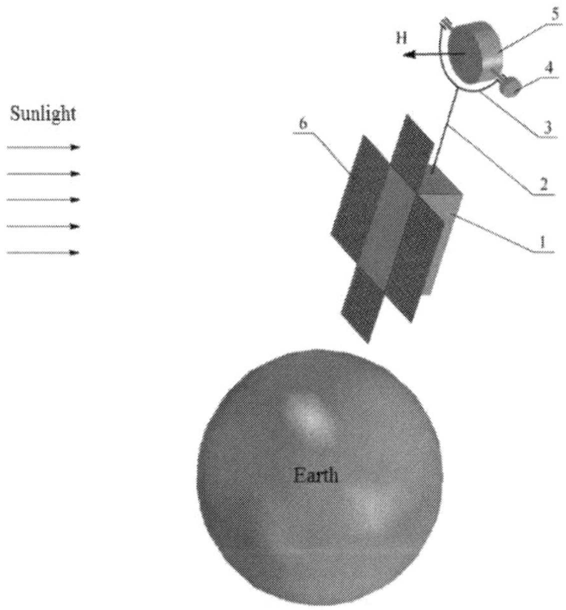

Figure 2. Kinematic scheme of the mechanical system "spacecraft – rod –gyrodyne".

Figure 2 shows the gravitational system of orientation, which consists of spacecraft (1); rod (2); suspension (3); torque sensor (4); gyromotor (engine-flywheel) (5); solar power panel (6).

Rod 2, which turns from the transport position to the working one, is fastened to spacecraft housing 1 (figure 2). The gyrodyne is fastened at the end of the rod and consists of gyromotor 5, frame 3 and torque sensor 4. In this case, the spacecraft will be stabilized along the roll and pitch positions by means of the gravitational moment from rod 2. The angle yaw relative to the axis of the rod provides the control of the angular position of the spacecraft via the gyroscopic moment when power is supplied to torque sensor 4.

The gyrodyne (figure 3) consists of housing 2 fastened to rod 1, wherein gyromotor 3 is attached by semi-axes 4. In the general case, the vector of the gyromotor kinematic momentum is directed perpendicularly to the orbit plane. Semi-axes 4 can rotate in housing 2 through bearing supports 6. Torque sensor 5 is attached to the semi-axes of the suspension element.

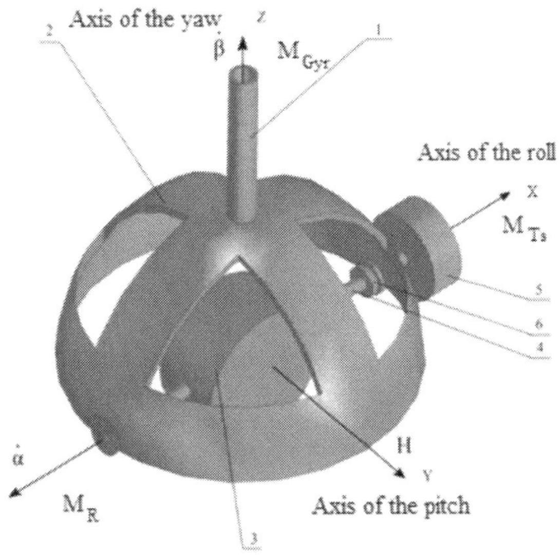

Figure 3. General arrangement of the gyrodyne.

When power is supplied to the sensor stator winding in accordance with the theory of gyroscopes under the law of the precession relative to the gravitational rod axis, the gyroscopic moment M_{Gyr} appears and this causes angular motion of the mechanical system "gyromotor– spacecraft – rod" relative to the axis of the rod with angular rate $\dot{\beta}$.

The presence of the angular rate $\dot{\beta}$ relative to the axis of the rod causes the moment M_R, which is equal to the moment M_{Ts}. The moment M_{Ts} is produced by torque sensors, and it is oppositely directed, i.e. $M_{Ts} = -M_R$.

Cutoff of power supply to the torque sensor via the control unit stops mechanical action on the spacecraft $\dot{\beta}$. In this case, stabilization of the spacecraft angular position is performed by the gravitational moment produced by the rod and kinematic momentum H of the gyroscope.

3.1 Principles of control

Attitude control can be fulfilled different ways [6].Three-axis control of the considered spacecraft is performed as follows.

Stabilization relative to the roll and pitch axes is achieved by the gravitational principle. Angle yaw control is performed under the law of precession as noted above. In addition, the feature of the gyrodyne is the variable speed of the gyromotor, i.e. the gyromotor is a cosmic engine-flywheel [7, 8].

Jaw control and realization of the required direction of the small spacecraft with adequate accuracy is performed as follows.

Figure 4 shows the area on the surface of the Earth divided into four quadrants. The objects to be scanned are dots 1 and 2. The scanning device is directed to dot “0” with the account of the precision of the gravitational system of orientation.

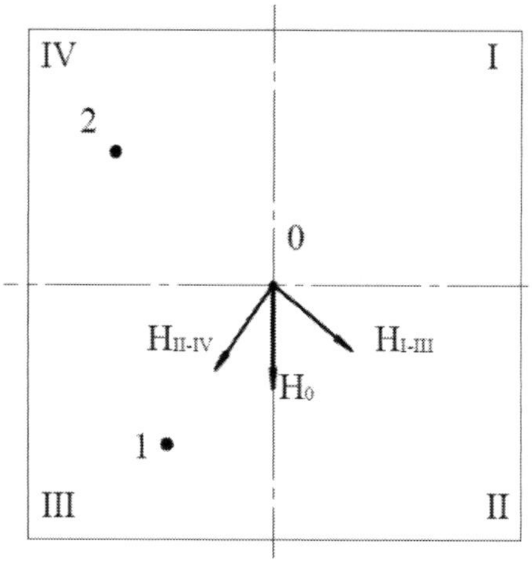

Figure 4. Graphical representation of the gravitational rod axis displacement on the Earth`s surface.

For targeting at dot 1, the torque sensor is turned on. It rotates the small spacecraft about the gravitational rod axis, and the momentum H of the gyrodyne moves from position H_0 to position $H_{I\text{-}III}$. As a result, dot 1 holds the position in the line, which is perpendicular to the momentum. After that, the regulated-velocity engine-flywheel is turned on (gyromotor 3). When electromechanical torque changes its magnitude, the equipment of the small spacecraft will be targeted at position 1, or any other position inside I or III quadrants.

To target the small spacecraft at quadrants II or IV, the momentum H is displaced from position H_0 to position $H_{II\text{-}IV}$. After that, the flywheel targets the device sat dot 2 or any other dot inside quadrants I or III.

When the small spacecraft, which is cylindrically symmetrical and has different moments of inertia, moves inside the central Earth`s gravitational field, itis affected by torques, which stabilize it in the Earth`s coordinate system (figure 5). These torques can be approximately determined by the following expressions:

in the yaw plane

$$M_z = \frac{1}{2}\omega_0^2 (J_x - J_y) \sin 2\psi \tag{1}$$

in the pitch plane

$$M_x = \frac{3}{2}\omega_0^2 (J_z - J_y) \sin 2\varphi \tag{2}$$

in the roll plane

$$M_y = \frac{3}{2}\omega_0^2 (J_z - J_x) \sin 2\vartheta \tag{3}$$

J_x, J_y, J_z are the main central moments of inertia of the small spacecraft;

ω_0 is the orbital rate of the small spacecraft;

ϑ, φ, ψ are the angles of pitch, roll and yaw.

These expressions show that restoring torque depends on the difference between the moments of inertia relative to the axes (direct relation), and this torque increases as a sinus of the double angle between the actual and null position of the gravitational rod axis.

3.2 Dynamics analysis of control

Full differential equations of motion of the mechanical system "small spacecraft – rod – gyrodyne" obtained using the second method of Lagrange are complex and interrelated.

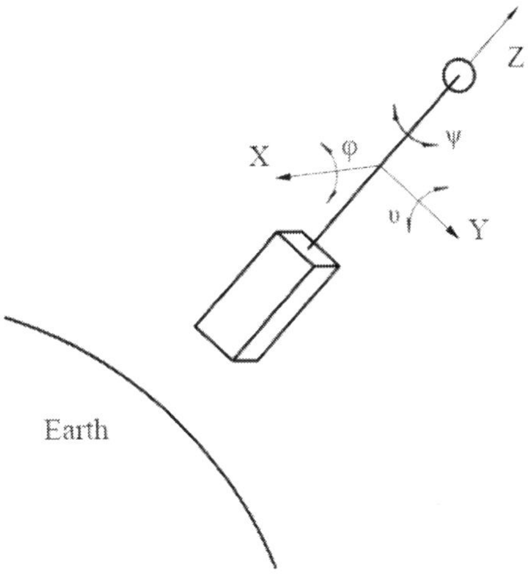

Figure 5. Schematic diagram of the gravitational system of orientation.

After separation from the mother missile, the spacecraft passes through a calm stage. It is affected by the gravity and gyroscopic moments and makes oscillations of small magnitude and rate. Therefore the roll, yaw and pitch positions of the small spacecraft, as well as the angular velocity with respect to these axes can be considered negligible, and the simplified equations take the form [4,9]:

$$J_x \cdot \ddot{\varphi} + (J_y \cdot \omega_0 - J_F \cdot \omega_F) \cdot \dot{\psi} + \left[3 \cdot (J_z - J_y) \cdot \omega_0^2 + J_F \cdot \omega_F \cdot \omega_0\right] \cdot \varphi + J_F \cdot \dot{\omega}_F \cdot \psi = 0 \quad (4)$$

$$J_y \cdot \ddot{\psi} + (J_F \cdot \omega_F - J_y \cdot \omega_0) \cdot \dot{\varphi} + \left[3 \cdot (J_z - J_x) \cdot \omega_0^2 + J_F \cdot \omega_F \cdot \omega_0\right] \cdot \psi + J_F \cdot \dot{\omega}_F \cdot \varphi = 0 \quad (5)$$

$$J_z \cdot \ddot{\vartheta} + \omega_0^2 \cdot (J_x - J_y) \cdot \vartheta + J_F \cdot \dot{\omega}_F \cdot \dot{\alpha} = 0 \quad (6)$$

$$J_e \cdot \ddot{\alpha} = M_G + J_F \cdot \omega_F \cdot \dot{\vartheta} + J_F \cdot \omega_F \cdot \dot{\psi} \quad (7)$$

J_x, J_y, J_z are the main central moments of inertia of the small spacecraft;
J_F is the momentum of inertia of the flywheel;
ω_F is the angular rate of the flywheel;
ω_0 is the angular orbital rate of rotation of the small spacecraft;
$\dot{\alpha}, \ddot{\alpha}$ are the angular acceleration and rate of the engine-flywheel in the uniaxial suspension element;
J_e is the equatorial moment of inertia of the flywheel.

To simplify the system of differential equations (2-7) and to gain better understanding of the dynamics of the mechanical system, some members are neglected. These are, for example, members that depend of damping properties of the rod material, construction material, active damping devices, which dissipate the kinetic energy of the rotational motion, and bending oscillations of the gravitational rod near null position.

Equations (2-7) show that the yaw motion of the small spacecraft in the orbital plane is not related to the roll and pitch motions. Application of the gyrodyne with the regulated-velocity engine-flywheel allows control along the yaw, roll and pitch axes and increases restoring torque.

4 Conclusion

The paper describes a hybrid gravitational orientation system of small spacecraft which uses a biaxial gyroscope with the regulated-velocity flywheel firmly fixed at the end of the gravitational rod as load. It significantly expands the functionality of this orientation system.

This system is hybrid as it combines the main advantage of the passive system (minimization of energy consumption during pauses in operation of the payload) and remains constantly directed towards the Earth's surface.

This gravitational orientation system is provided with software to perform targeting during operation with adequate accuracy. The accuracy of the gravitational system of orientation is about $(5-10)\,^{\circ}$, while the hybrid system in the active mode can realize the accuracy within arcminutes.

Since the equipment does not work all the period of the aircraft operation, energy is saved, and this is crucial for small spacecrafts.

Moreover, this orientation system has an operational advantage: replacement of the passive load to an active control element of equal mass.

References

1. Y. Wang, S. Xu, J. Aerospace Eng. **27**, 141 (2014)
2. S. A. Rawashdeh, University of Kentucky Master's Theses, 624 (2010)
3. A. M. Si Mohammed, M. Benyettou, Y. Bentoutou, A. Boudjemai, Y. Hashida, M. N. Sweeting, Acta Astron. **64**, 802 (2009)
4. H. J. Fletcher, L. Rongved, E. Y. Yu, Bell Syst. Tech. J. **42**,2245 (2013)
5. C. Circi, J. Aerospace Eng. **221**, 87 (2007)
6. Godard, K. D. Kumar, Guidance, Navigation, and Control Conference, 8197 (2010)
7. D. J. Richie,V. J. Lappas, B. Wie, ASCE, 7501 (2008)
8. H. Yoon, P. Tsiotras, J. Guid. Control Dyn. **25**, 1084 (2002)
9. T. Pulecchi, F. Casella, M. Lovera, Simul. Model. Pract. Th. **18**,77 (2010)

MATEC Web of Conferences 48, 01003 (2016)
DOI: 10.1051/matecconf/20164801003

Numerical computation of the main gas-dynamic bearing static characteristics for the ball gyroscope

Anastasiya Ignatovskaya[1,a] and Alexey Golikov[1]

[1] National Research Tomsk Polytechnic University, 634050 Tomsk, Russia

Abstract. The numerical computation questions and analysis of the engineering calculation of the main gas-dynamic bearing static characteristics for the ball gyroscope experimental model are considered. The developed gyroscopic device construction is provided. These investigations for evaluation of the possibility to realize a sensitive element, which base on the developed device were provided. It is supposed, that this sensitive element can be used in information measuring systems of various applications. The main calculation data, such as load-bearing capacity and other characteristics of the hemispherical gas-dynamic bearing is carried out. According to the attained results, preliminary implications are stated. The basic problems for the further research are determined.

1 Introduction

One of the most important gas bearing advantages is using the ambient air as a lubricant reservoir directly. A bearing, where gas is injected in the lubricating gap by the motion of the lubricated surfaces and there are no additional sources of pressurized gas, are known as self-supporting or gas-dynamic. This type of bearing was taken as a basis for the designed device [1].

Gas lubrication has special features which are typical for its nature. In comparison with liquids, gas has the less viscosity. Ambient temperature acts on the gas properties with a small effect. Ambient pressure impacts on the gases far less. Such viscous stability and its small quantity open the wide variety of gas bearing application in devices working on a high speed with a wide range of operational temperature and pressure [1]. Therefore, gas bearing application is determined by these unique advantages in cases when the traditional types of bearings lose their properties [2].

At this stage, the main project purpose is to define the mathematical model for calculation of the main static gas-dynamic bearing characteristics. It is necessary for evaluating the possibility to use this kind of bearing as a suspension of the gyroscope ball rotor. It is planned that developed gyroscopic device can be used as a position sensor of information measuring systems working under the severe climatic and mechanical conditions. Due to this fact, such sensor has to meet requirements connected with efficiency, surety, and accuracy [1, 3].

[a] Corresponding author : nastena@tpu.ru

2 Device construction

Figure 1 shows the developed gyroscopic device construction with gas-dynamic bearings. The main elements of the experimental model are: as the bearings, there are two hemispherical bowls 7, rotor 1 is placed in the cavity of these bowls. It is a standard ball bearing. It has an axial bore, additionally. Moving parts of the dual axis angle transducer 4 are pressed in the pole places of the axial bore. The transducer mating part 3 is placed in one of the bowls. The diameter of the bowl forming sphere for 5÷10 mm more than actual diameter of the ball rotor, which is 28,587 mm. Thereby, such difference provides the initial gap which is necessary for operational regime.

The ball rotor is rotated by a three-phase asynchronous motor 2. Power 40 V, at frequencies 500 or 1000 Hz.

In general view control position elements of the ball rotor are not shown.

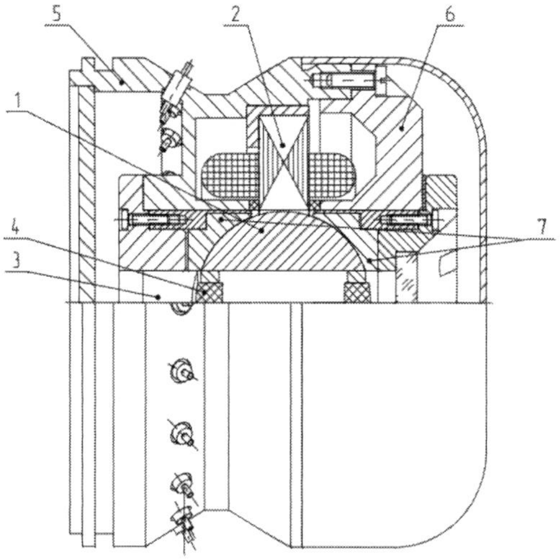

Figure 1. General view of the gyroscope experimental model:
1 –ball rotor, 2 – stator part, 3 – dual axis angle transducer, 4 – moving part of the transducer, 5 – case, 6 – cover, 7 – hemispherical bowls.

3 Modeling procedure

Experimental and, especially, theoretical research plays a crucial role in assessing the gas-dynamic bearing characteristics. Mathematical modeling comes to the fore as an experiment for units operating on gas lubricant at the design stage. It is attributable to difficulties and expensiveness of the high precision manufacturing suspension parts.

The principal characteristics estimating efficiency and reliability of the units with gas lubrication are: load-bearing capacity, bearing stiffness, and quantities of the viscous and dry friction momentums (the last value is important only for the initial start time) [3].

A number of several characteristics such as bearing radius, clearance, presence of the grooves, parameters of the gaseous medium (viscosity, free-path length of gas molecules, pressure and temperature), magnetic traction, geometric errors of the “contact” surfaces, injection capability, bearing microprofile, and other parameters, have a huge impact on the gas-dynamic bearing characteristics in general [4].

At this point, there are a number of universal software tools using to solve some problems of gas dynamics. These software tools are: LS-DYNA, ABAQUS, STAR-CD, ANSYS modules - CFX and Fluent, Flow Vision and others [5].

Among them, the most appropriate software package is ANSYS 15.0 (Fluent and CFX modules). The mathematical model is based on solving the equations system. This system consists of the fundamental laws of mass, momentum and energy conservation. The system is closed by initial and boundary conditions, as well as defining relations. If some effects are not accounted by the grouped equations, it will taking into account by the special turbulence equations. The resulting synthesized system is the Navier-Stokes equation. It is general equation of the laminar flow of viscous gas dynamics.

Initial data for the computation is shown in Table 1. It contains information about geometric and gaseous medium parameters, and rotation speed.

Table 1. Initial data for parameters computation of the working gas-dynamic bearing variant (unshaped).

Mathematical model	AKN, SST
Gap range, microns	2÷20
Gap minimum value providing by the existing elements, microns	5
Rotor nominal speed at the supply frequencies, rad/s 500 Hz 1000 Hz	 850 1675
Gaseous medium	air
Gaseous medium model	incompressible ideal gas
Temperature and pressure conditions	normal

Using software package ANSYS 15.0 – Fluent, two-dimensional model of the system "rotor-bearing" was formed. Also, operating pressure in the gap bearing was calculated. The simulation results are presented in Figure 2.

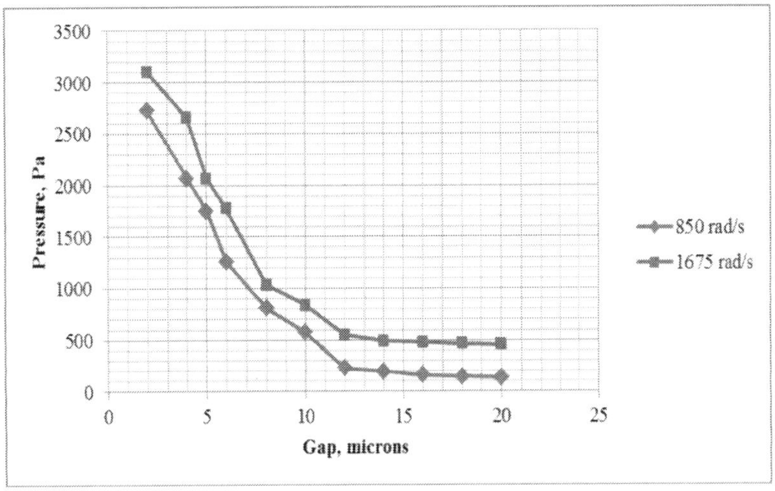

Figure 2. Dependence of the pressure distribution on the gap bearing at the rotor nominal speed 850 rad/s and 1675 rad/s.

If clearance increases, pressure in the gap will decrease. It reduces load-bearing capacity efficiency. The highest pressure can be obtained when the gap magnitude less than 2 microns.

In reality, the minimum gap value, which can be provided by existing elements base of the experimental model, is about 5 microns.

Table 2 presents mathematical modeling results and estimating computation of the gas layer parameters at the 5 microns gap. It was provided at two basic nominal frequencies of rotation as shown in Table 1.

Table 2. The main gas-dynamic bearing characteristics (unshaped).

Parameter	**Value**	
Rotor speed, rad/s	850	1675
Ball rotor weight, N	0,8	
Reference suspension area, cm^2	4	
Operating pressure, Pa	1749,2	2067,3
Load-bearing capacity, N	0,7	0,83
Maximum viscous momentum, N·m	0,0012	0,0023

Calculated values of the load-bearing capacity show the following. When the gap magnitude is equal to 5 microns and rotor speed is 850 rad/s, bearing, at a given load (own weight), will not work in the gas lubrication regime. The load-bearing capacity is less than the ball rotor weight.

In comparison with the full-scale simulation, results are different. It is assumed, that geometric and technological errors of the hemispherical bowls working surfaces increase the injection capability. In theoretical model such results were not observed, since the injection capability was not included, due to the consideration of smooth, perfectly shaped hemispherical bowls [6].

At the rotor speed 1675 rad/s, bearing worked in the gas lubrication regime, which was confirmed experimentally.

The viscous friction momentum effects on the rotor acceleration, determines the required electric power, characterizes heating, and influences on the device sensitivity. The more the gap bearing is, the less viscous friction momentum effects. At the same time, the load-bearing capacity and rigidity become lower, and device sensitivity decreases too [6]. Decreasing of the last values will adversely affect the possibility of using this type of bearing as the suspension for the gyroscopic sensor. Therefore, the main solution for reducing viscous friction momentum influence is to feed into gaseous medium (not air). It will reduce the aerodynamic drag and accelerate the processes of heat removal, as well as allows designing of a reliable sensory system with satisfactory accuracy [7].

4 Conclusion

Investigations of the main gas-dynamic bearing static characteristics for the ball gyroscope show the convergence results for both mathematical and physical experiments [6]. It is found that the effectiveness of this bearing realization is not high in terms of load-bearing capacity and, obviously, the device sensitivity as an inertial sensor. It should be noted that geometric errors of the rotor surfaces and hemispherical bowls were not taken into account in the theoretical model. Further research will be focused on improving physico-mathematical model, parameters optimization [8], as well as to address the other issues related to costing and achieving acceptable results.

References

1. I.E. Sipenkov, A.Yu. Filippov, Yu.Ya. Boldyrev, B.S. Grigoriev, N.D. Zablotchkiy, G.A. Luchin, T.V. Panich, *Precizionnye gazovye podshipniki* (Electropribor, St-Petesburg, 2007)
2. I.V. Yakovlev, *Issledovanie vibracij, prochnosti i konstrukcii detalej aviadvigatelej* (Mashinostroenie, Moscow, 1968)
3. A.N. Golikov, A.A. Ignatovskaya (Kuzma), M.S. Plehanov, *Jelektronnye i jelektromekhanicheskie sistemy i ustrojstva* (AO NPC Poljus, Tomsk, 2013)
4. A.V. Dubinin, K.V. Smolyan, Prib. **S5**, 169 (2012)

5. Y.M. Temis, M.Y. Temis, A.B. Meshcheryakov, J. Frict. and Wear **32(3)**, 212 (2011)
6. A.A. Ignatovskaya, A.N. Golikov, J. Phys.: Conf. Ser. **671**, 012020 (2016)
7. V.I. Grabovskij, Izvestiya Akademii Nauk. Mekhanika Zhidkosti/ Gaza **2**, 33 (2003)
8. E.A. Dukhovskoi, L.V. Voitenko, A.V. Lyubavin, B.M. Krivitski, Trenie. Iznos **2(6)**, 1099 (1981)

MATEC Web of Conferences **48**, 01004 (2016)
DOI: 10.1051/matecconf/20164801004

Analytical studies of torque motor tape active element

Antonina Dolgih[1,a], Vladimir Martemyanov[1] and Ivan Samodurov[1]
[1]National Research Tomsk Polytechnic University, 634050 Tomsk, Russia

Abstract. The paper presents analytical studies of the torque motor tape active element. The tape active element is a novel type of a motor's stator organization, where the conventional winding is replaced by a tape winding. Given the operation principle of proposed active element; its torque characteristics are then computationally found with using the finite element method (FEM). The results show the possibility of the optimal value of the relative electrode width, when the torque will be maximal. The analytical studies of the motor's torque over the number of tape winding coils allowed to receive the recommendations on choosing the number of coils.

1 Introduction

Recently, the significant attention is given to the development of the brushless torque motors novel types. The execution variant of the novel type torque motor is proposed in [1, 2]. In the active element of this motor the conventional winding is replaced by a tape winding. The tape winding represents a spirally wound tape on a tubular base. The tubular base is a part of the motor case. The tape is made of electroconductive material and coated with a layer of insulation material. At both side edges of the tape there are narrow transverse cuts. The distance between cuts centers is equal to the pole division of the rotor magnetic system. If the ends of the wound tape connect to the DC source, the character of current flowing is determined by the transverse cuts. In the Figure 1 [1] it is shown that the current J has two components: the longitudinal one J_l directed along the tape and the transversal one J_t directed across the tape towards coinciding with the rotation axis of the torque motor rotor. The poles of the rotor magnet are displayed by rectangles N and S. The interaction of current transversal components J_t with the magnetic field causes forces F, acting on the magnet. The total action of these forces causes the torque that tends to turn the magnetic system relative of the stationary tape winding. The longitudinal current components J_l cause the forces P, acting from the side of magnet to the rotor suspension supports.

The benefits of the proposed active element include:

- Novel design and technological approach to the engine stator manufacture
- Great linear load due to the high current flowing through the tape winding and a good heat removal to the motor case
- Possibility of formation the required functional dependence of the motor's torque over the angular movement characteristic

[a] Corresponding author : ivanovatonya@tpu.ru

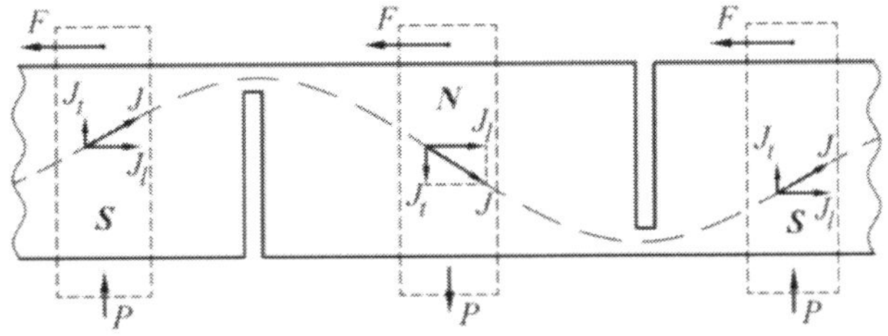

Figure 1. The active tape element fragment.

The operating principle of the proposed active elements is described in [2] by the example of the diagonally placed thin current-carrying conductor and magnetic field interaction. It is known that the diagonal character of the current flowing is also implemented in the windings of the coreless motors. Such windings are known as honeycomb, winding with skew-wound coils, zig-zag windings, Faulhaber winding [3-5].

The coils of the honeycomb winding are placed close or alternate to each other at an angle to the rotor element of cylinder. From design point of view the proposed variant of the active element is close to the approach implemented in the honeycomb winding. The main benefits of honeycomb windings include [5]:

- No cogging torque resulting in smooth positioning and speed control and higher overall efficiency than other DC motor types
- Extremely high torque and power in relation to motor size and weight
- Absolute linear relationship between load to speed, current to torque, and voltage to speed
- Very low rotor inertia which results in superior dynamic characteristics for starting and stopping
- Extremely low torque ripple and EMI

Torque motor with the tape active element has the dependence of motor's torque over the angular movement shown in the Figure 2, curve 1.

This dependence is obtained by calculation method [6] with using a numerical simulation of the current distribution through the tape winding in COMSOL Multiphysics. COMSOL Multiphysics software is a universal computer aid engineering software based on finite element analysis, which has a large set of functions for analyses and solution. For calculation of the dependence, the tape winding consists of 100 coils with 4 plates in each coil. Plate's length is 100 mm, plate's width is 50 mm and plate's thickness is 0,1 mm. Current is 50 A, flux density in the air gap is 0,5 T. Curve 2 in the Figure 2 represents the dependence of the motor's torque over the limited angular movement with the additional side cuts in the tape [6]. The additional side cuts allows to obtain the steady torque over the angular movement.

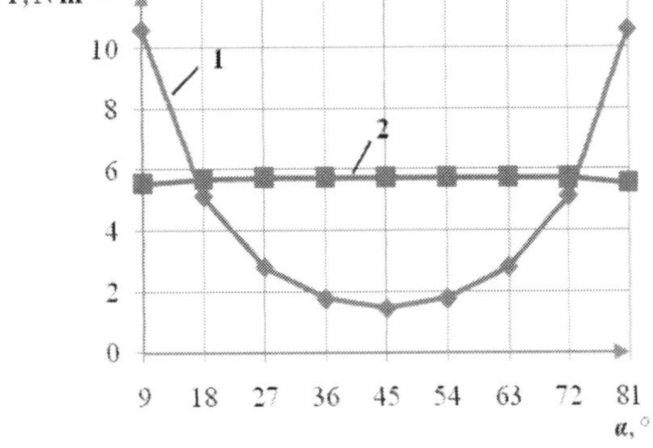

Figure 2. Calculation result of the motor's torque over angular movement.

The purpose of this work is to carry out the analytical studies of the tape active element single plate geometry influence and the number of tape winding coils on the motor's torque by the numerical simulation in Comsol Multiphysics.

2 Results and discussion

As depicted in the Figure 1, the distributed direct current J flows through the side electrodes of the single plate in the diagonal direction. Motor's magnetic system provides the uniform magnetic field with the constant flux density for some area of plate. The magnet center situated at some fixed point x_0 relative to the plate. Taking into account that the transverse current density component j_y has the same value for the points with the same coordinates (x_i, y_i), situated in all plate layers. The plate thickness Δ is constant. Suppose that the plate temperature, the applied voltage and the conductivity of the plate material are constant in time. In practice it is possible with the short time current flow, for example at the impulse mode.

In [6] it is shown that force generated by the interaction between the current-carrying plate and the permanent flux density has the following form

$$F_x = B \cdot \Delta \cdot \int_{(x-c/2)}^{(x+c/2)} \int_0^b j_y(x, y) \cdot dx \cdot dy, \tag{1}$$

where B is a flux density in the air gap,
c is a magnet width,
b is a plate width.

The double integral in (1) represents the sum of the transverse current density components, flowing through the plate. Introduce the notation:

$$D(x, y) = \int_{(x-c/2)}^{(x+c/2)} \int_0^b j_y(x, y) \cdot dx \cdot dy. \tag{2}$$

If the value of $D(x, y)$ with some initial current $J = J_0$ is known, the double integral is denoted as $D(x, y, J_0)$. With current $J \neq J_0$ we have

$$D(x, y, J) = \frac{J}{J_0} D(x, y, J_0). \tag{3}$$

In this case, (1) can be written in the form

$$F_x = \frac{B \cdot \Delta}{J_0} \cdot J \cdot D(x, y, J_0). \tag{4}$$

The motor's torque is determined by the combined action of the tape winding single plates.

2.1 The active element geometry influence to the motor's torque

Consider the tape active element single plate geometry influence on the motor's torque. Firstly assess the force behavior over the changing of the voltage application electrode relative width $\sigma = \frac{b_e}{b}$, where b_e is the electrode width (Figure 3).

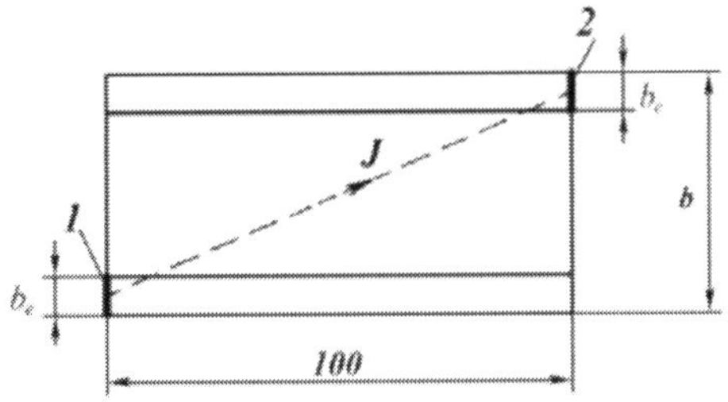

Figure 3. The single plate of the tape active element.

The operating current value J at a constant voltage U, applied to the plate between the electrodes 1 and 2 depends on the relative electrode width σ. This occurs because the changing of the electrode cross-section area causes the change of the plate resistance and operating current.

The numerical experiment under specific values of σ, c, x_0, J_0 was carried out. In the Figure 4 the graph $J(\sigma)$ is represented. There is σ varies from 0,1 to 0,5 with the constant voltage, in particular $U = 0{,}01$V.

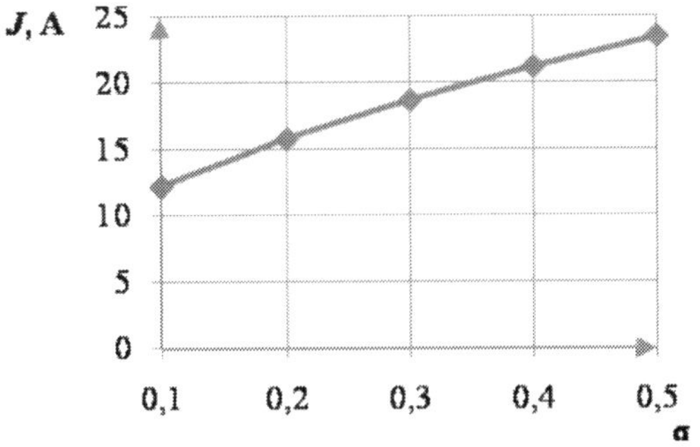

Figure 4. The graph of operating current over the relative electrode width with the constant voltage.

Note that the value $D(x, y, J_0)$ depends on the current streamlines steepness. The Figure 5 shows that increasing of σ decreases the steepness.

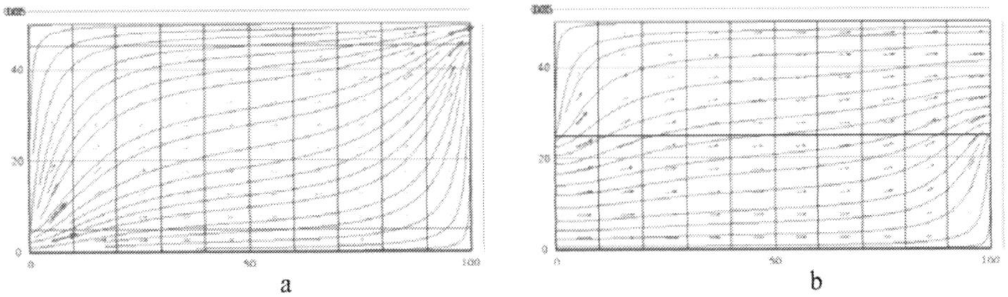

a b

Figure 5. Current streamlines: a when $\sigma = 0{,}1$; b when $\sigma = 0{,}5$.

Define the value of $D(x, y, J_0, \sigma)$ with σ from 0,1 to 0,5 (Figure 6). For providing the constant current J_0, flowing through the plate, in particular 1 A, the applied voltage should be changed each time. The curve 1 shows the magnet location at the edge of the plate, curve 2 - at the center of the plate.

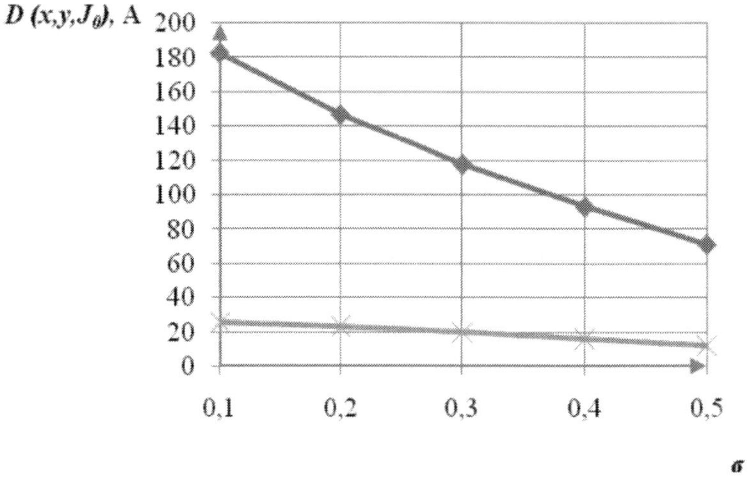

Figure 6. The double integral over the relative electrode width with the constant current J_0 .

Thus the increasing of the σ causes the increase in the current J (Figure 4) and the decrease of the $D(x, y, J_0)$. Consequently, the simultaneous action of two opposite trends, affecting on the force value is observed. According that, (4) can be written as

$$F_x(x_0) = \frac{B \cdot \Delta}{J_0} \cdot J(\sigma) \cdot D(x, y, J_0, \sigma). \tag{5}$$

The resulting force is the product of the above functions and has the form shown in Figure 7. Due to the fact that the both functions have a non-linear character, the expression for the force has an extremum at a certain value of the relative electrode width.

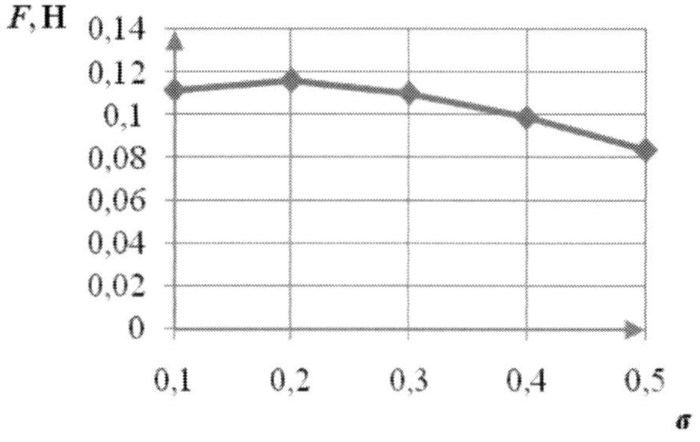

Figure 7. The resulting force over the relative electrode width.

2.2 The motor's torque over the number of tape winding coils

Consider how the motor's torque depends on the number of tape winding coils N. In this case, (4) takes the form

$$F = B(\delta) \cdot \frac{\Delta}{J_0} \cdot \frac{U}{R \cdot N \cdot 2p} \cdot D(x, y, J_0)_{x_c}, \tag{6}$$

where $B(\delta)$ is a flux density in the air gap; $\dfrac{U}{R\cdot N\cdot 2p}$ is the tape winding operating current; $U = const$ – the voltage, applied to the tape; R is the resistance of the single plate; $2p$ is the number of plates in one coil; $D(x,y,J_0)_{x_0}$ is the double integral defined with the current J_0, and with the magnet center in the point x_0.

The force, created by the pack from N one pole plates has the form

$$F_N = B(\delta)\cdot\frac{\Delta}{J_0}\cdot\frac{U}{R\cdot N\cdot 2p}\cdot N\cdot D(x,y,J_0)_{x_0}. \quad (7)$$

The force, created by the all $2p$ poles

$$F_{N2p} = B(\delta)\cdot\frac{\Delta}{J_0}\cdot\frac{U}{R\cdot N\cdot 2p}\cdot N\cdot 2p\cdot D(x,y,J_0)_{x_0} = B(\delta)\cdot\frac{\Delta}{J_0}\cdot\frac{U}{R}\cdot D(x,y,J_0)_{x_0}. \quad (8)$$

Then the motor's torque can be defined by the following way

$$M = F_{N2p}\cdot r_{cp} = B(\delta)\cdot\left[\frac{r_{av}\cdot\Delta}{J_0}\cdot\frac{U}{R}\cdot D(x,y,J_0)_{x_0}\right], \quad (9)$$

where r_{av} is the average radius of the tape winding.

From (9) follows that the torque value depends only on flux density, and consequently from the value of the air gap. On the one hand the increased number of coils should increase the torque, but at the same time the current is reduced. Moreover, the larger number of coils will increase the air gap and decrease accordingly the flux density. Therefore, the number of coils must be minimal and defines by a permitted current. The permitted current J_{per} must not damage the power supply and the tape winding

$$\frac{U}{R\cdot N\cdot 2p}\le J_{per},\ N\ge\frac{U}{2p\cdot R\cdot J_{per}}. \quad (10)$$

Provided that the magnetizing force of the permanent magnets is constant, we have

$$B(\delta) = a\cdot\delta^2 + b\cdot\delta + c = a\cdot\left(\delta^2 + \frac{b}{a}\cdot\delta + \frac{c}{a}\right). \quad (11)$$

Assume the air gap value in the form of

$$\delta = N\cdot\Delta + \Delta_m, \quad (12)$$

where Δ_m is the mechanical gap, including the thickness of the tubular base.

Substituting (11) and (12) into (9), we obtain

$$M = k\cdot\Delta^2\cdot\left[N^2 + B\cdot N + C\right] \quad (13)$$

where $k = a \cdot \left[\frac{r_{av} \cdot \Delta}{J_0} \cdot \frac{U}{R} \cdot D(x, y, J_0)_{x_0} \right]$, $B = \frac{1}{\Delta}\left(2 \cdot \Delta_m + \frac{b}{a}\right)$, $C = \frac{1}{\Delta^2} \cdot \left(\Delta_m^2 + \frac{b}{a} \cdot \Delta_m + \frac{c}{a}\right)$.

Based on the possible application areas of the torque motors, among the basic requirements, applied to all the torque motors [7], it should be emphasized that the ratio of torque to the volume, weight, power consumption, management power should be higher to provide the quality of the motor. The power consumption defines by the following way

$$P = J \cdot U = \frac{U^2}{2p \cdot N \cdot R}. \tag{14}$$

Define the torque-power ratio

$$\frac{M}{P} = \frac{k \cdot \Delta^2 \left[N^2 + B \cdot N + C\right]}{U^2} \cdot 2p \cdot N \cdot R. \tag{15}$$

Substitute the coefficient k into (8)

$$\frac{M}{P} = \frac{a \cdot \Delta^3 \cdot r_{av} \cdot 2p}{J_0 \cdot U} \cdot D(x, y, J_0)_{x_0} \cdot N \cdot \left[N^2 + B \cdot N + C\right] = K \cdot N \cdot \left[N^2 + B \cdot N + C\right] \tag{16}$$

After the simultaneous analysis of two expressions (13) and (16), taking into account (10), the number of coils *N*, which provide the characteristics of the motor defined by the customer determines.

3 Conclusion

The analytical studies of the tape active element single plate geometry influence on the motor's torque showed the possibility of the optimal value of the relative electrode width, when the torque will be maximal. The analytical studies of the motor's torque over the number of tape winding coils allowed to receive the recommendations on choosing the number of coils. A certain number of coils N will provide the motor characteristics defined by the customer.

References

1. V. Martemjanov, A. Dolgih (Ivanova), RF2441310 (2012)
2. V. Martemjanov, A. Dolgih (Ivanova), I. Siberian Conference on Control and Communications (SIBCON), 1 (2013)
3. R. D. Sedgewick, US4331896 A (1982)
4. J. F. Gieras, *Permanent Magnet Motor Technology: Design and Applications* (2010)
5. Dr. Fritz Faulhaber GmbH & Co. KG, *Technical information* (2016)
6. A. Dolgih (Ivanova), V. Martemjanov, I. Plotnikova, Instruments and Systems: Monitoring, Control, and Diagnostics **4**, 22 (2013)
7. L. Stolov, A. Afanasyev, *Momentnye dvigateli postojannogo toka* (1989)

MATEC Web of Conferences **48**, 01005 (2016)
DOI: 10.1051/matecconf/20164801005

Integrated approach of the dynamic characteristics optimization for spacecraft's technical systems on the exploratory tests

Victor Dmitriev[1,a] and Yuliya Britova[1]

[1] *National Research Tomsk Polytechnic University, 634050 Tomsk, Russia*

Abstract. Integrated approach is the process connecting various project, theoretical and testing procedures. It is provided by the specific series-parallel combination of the mentioned operations. This paper shows that using results of the previous procedures on the further algorithm stages of exploratory tests and product verification is expedient. At that, the necessary correcting of engineering solutions in corresponding procedures is conducted simultaneously. Then, results of each procedure are used in the fixed sequence. It provides maximum attainable optimality of the designed system.

1 Introduction

This paper presents the dynamic characteristics optimization method of complicated electromechanical system. This method is based on the integrated approach and combines results of analytical design and experimental verification. The suggested method of the integrated approach application in the sophistic technical systems designing such as final controlling element, which based on the torque-controlled reaction wheel motor (RWM), consists of the three stages (Figure 1) [1]:

1. Analytical design;
2. Experimental verification of an engineering sample;
3. Results analysis and design documents correcting.

All of these stages are visualized by the processes in flow chart. These processes are connected by the functional relations. The flow chart demonstrates the sequence technology of including the necessary engineering solutions correcting in providing processes at the given moment.

2 Methods. Experimental Procedure

The analytical design stage is based on the main operational and weight-size parameters, perturbed oscillations, and critical velocity calculations. Also, this stage includes parametric models design, characteristics optimization by the design requirements variation, and calculation of the element`s natural frequencies by using special computer-aided systems.

Experimental verification is essential continuation of the reaction wheel motor design. However, this procedure is not only the final part, but closely connects with the other design stages [2, 3].

[a] Corresponding author : dmitriev@tpu.ru

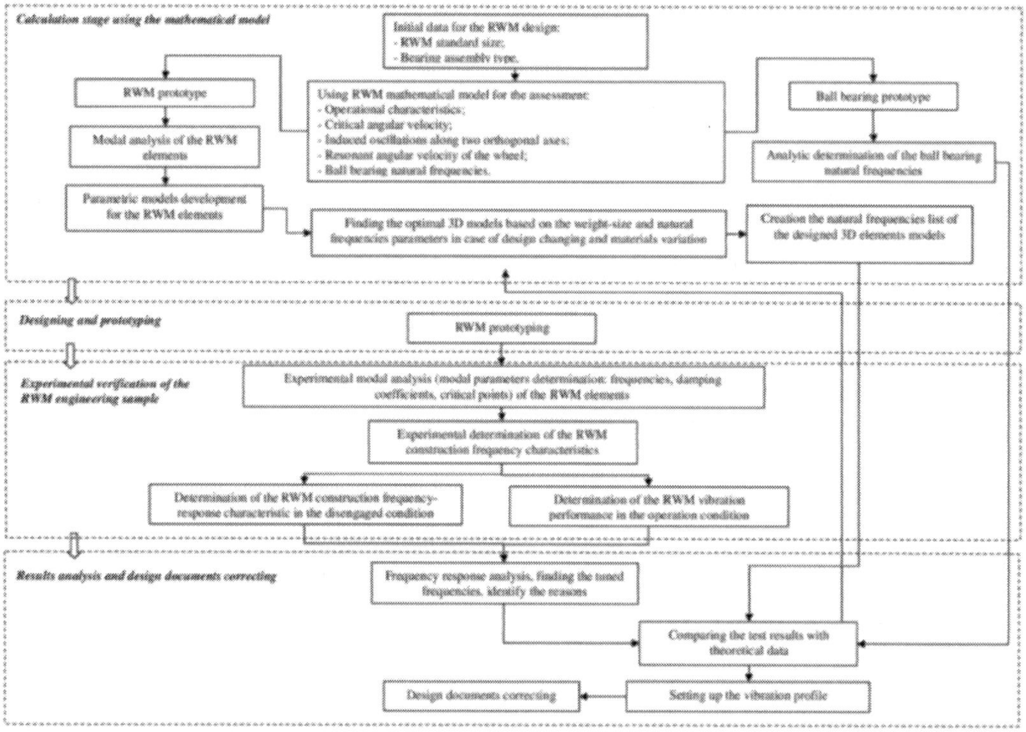

Figure 1. Block scheme of the integrated approach implementation.

Experimental methods for the determination causes of the RWM perturbation actions are based on the highly informative control procedures. At that, control and testing equipment are upgraded with the modern measuring systems based on existing diagnostic and test complex.

In the experimental RWM construction verification the following test types are used:

- Determination of the RWM vibration performance in the operation condition;
- Determination of the RWM tuned frequencies in the disengaged condition under the impact of sine wave vibration generated by the vibration system;
- Experimental modal analysis.

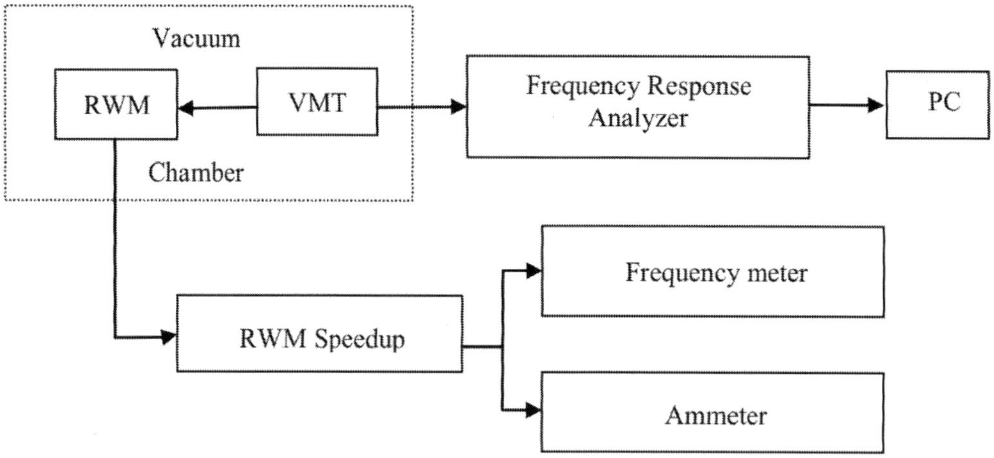

Figure 2. Flow chart of the RWM vibration performance determination: VMT – vibration-measuring transducer; PC – personal computer.

These testing approaches base on the modern non-destructive testing methods.

Method for the determination of the RWM construction vibration spectrum in the operation condition based on the vibration level from the construction elements during flywheel speedup and coasting. It can be provided in the angular velocities work range by means of the precision vibration-measuring transducers (VMT) [4].

The flow chart of the RWM vibration performance determination is shown in Figure 2.

In general, the RWM vibration performance is represented in Figure 3.

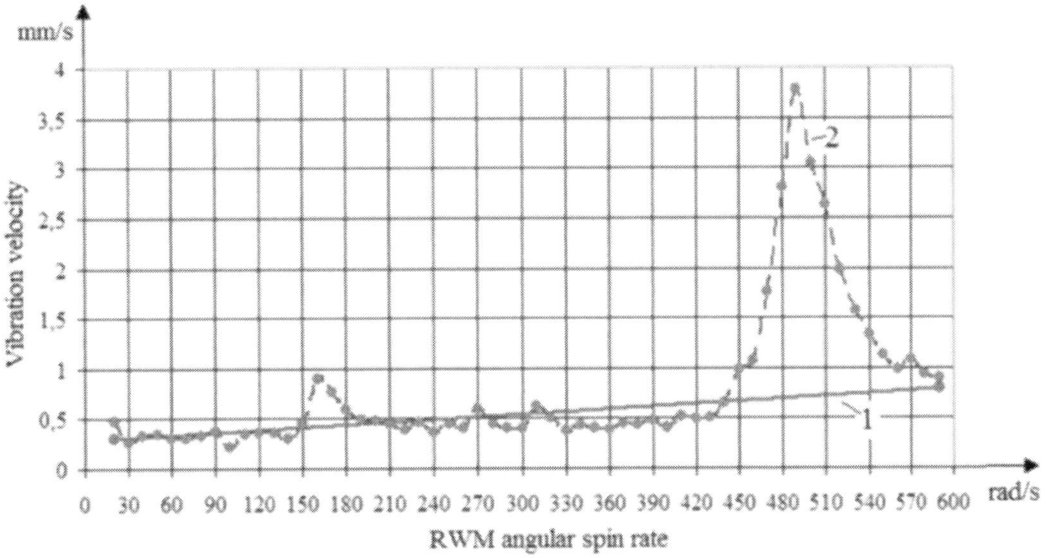

Figure 3. RWM vibration performances: 1 – Ideal; 2 - Investigated RWM.

Analysis and comparison of the obtained data with analytical modeling results and mathematical calculation of the ball bearing natural frequencies (the first stage of the integrated approach) allow identifying the causes of resonance [5, 6].

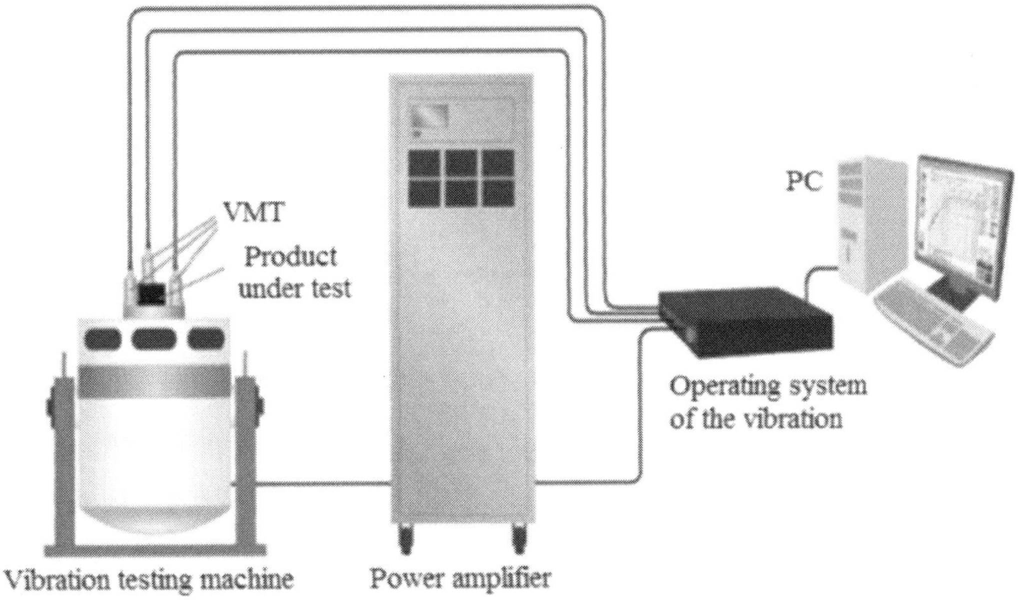

Figure 4. Test setup for the RWM tuned frequencies determination.

Identification methodology of the RWM tuned frequencies in the disengaged condition based on the determination frequency response from the RWM construction in three orthogonal axes with normalized impact which generated by vibration system. Test setup for the RWM tuned frequencies determination is illustrated in Figure 4.

As the test result the frequency-response characteristic (FRC) will. It is a graph of acceleration as a function of disturbing frequency.

This frequency-response characteristic can presents the tuned frequencies [7]. Resonant frequency is the frequency where the acceleration amplitude at the measurement point of FRC at two or more times greater than the acceleration amplitude of other frequencies in the spectrum. As an example, Figure 5 shows the RWM`s FRC along X-direction.

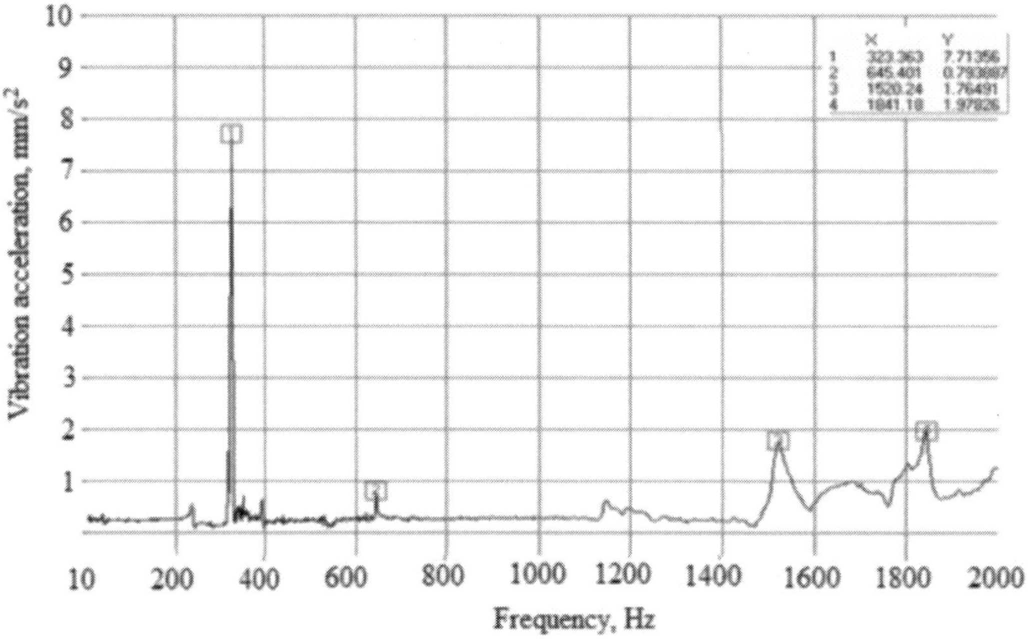

Figure 5. The RWM frequency-response characteristic along X-direction.

The modal parameters determination method (natural frequency, damping coefficient) (the flow chart is represented in Figure 6) of the RWM construction and its elements by the experimental modal analysis based on the calculating frequency response function for a point under the impact by a modal hammer or vibrator.

The experimental modal analysis allows conducting the investigations not only the entire product but also its individual elements [8].

The RWM experimental modal analysis example with using of the impact hammer is shown in Figure 7(a) and with the modal vibrator - Figure 7(b).

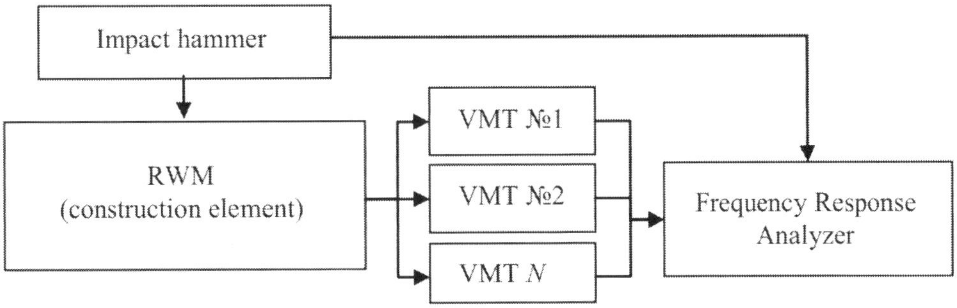

Figure 6. Test Flow Chart for the modal parameters determination.

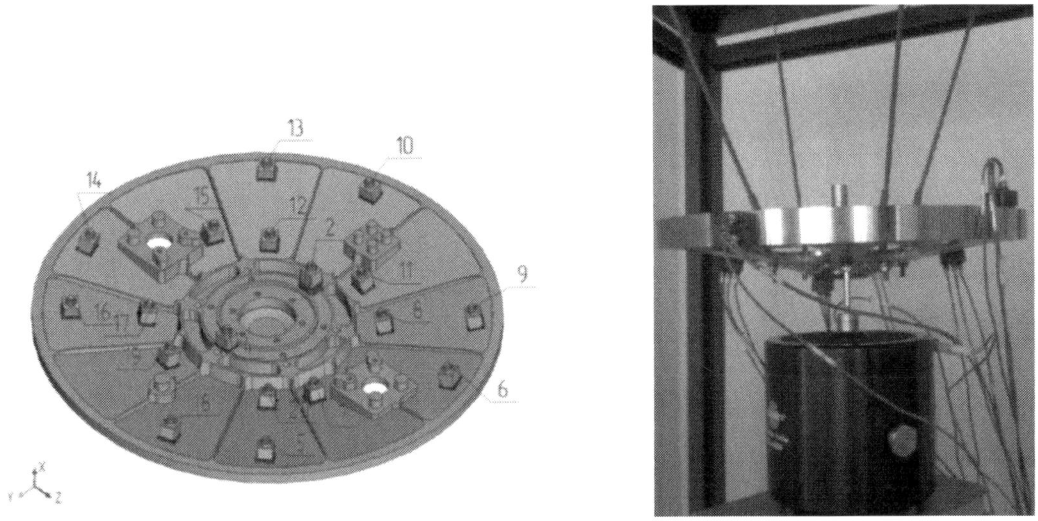

a b

Figure 7. Sensors arrangement and force application points on the RWM casing and base.

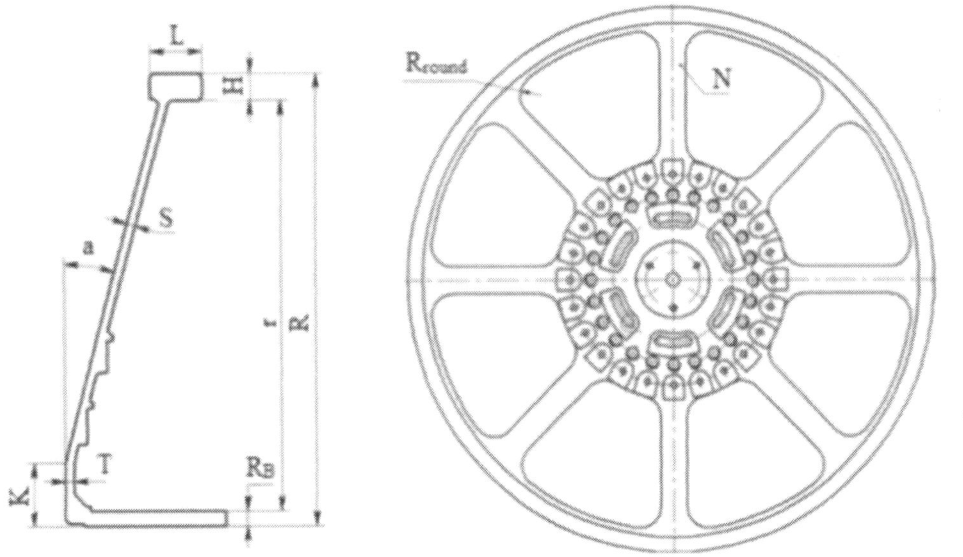

Figure 8. Structural dimensions of the flywheel model.
R - flywheel outside radius; r - flywheel inside radius; L - rim component width; H - rim component height; a - umbilical angle; S - umbilical thickness; R_B –shaft radius; T - space thickness for the fixing shaft; K - width of the space for fixing shaft; R_{round} – rounding radius of the umbilical spokes; N - number of spokes.

These frequencies can be compared with the values, which were determined by the modal frequency analysis in a computer-aided design. Frequencies which were identified by a special detail modal analysis can be detected in the RWM`s FRC. In addition, their contribution to the resonance distribution can be assessed. For increasing the "critical" elements, rigidity is necessary to apply to the first stage of the integrated approach and implement the parametric correcting of the selected 3D model construction element.

For instance, the RWM flywheel parametric 3D model is represented in Figure 8 [9, 10].

3 Conclusion

Analysis of the RWM vibration performance which was produced during speedup and coasting allows evaluating the impact of the RWM ball bearing natural frequencies on its resonance. Also, it leads to consider the possibility to correct axial load bearing or variation of the radial clearance.

This integrated approach provides developing the RWM construction meets the modern requirements.

The universal character of the developed method expands the number of application opportunities for designing in any complex technical systems.

References

1. Yu. A. Britova, *Thesis work,* 12 (2012)
2. S. Mohamady, A. Montazeri, R. Zahari, N. Aswan Abdul, *Advances in Acoustics and Vibration* **9**, 36 (2009)
3. Yu. A. Britova, V. S. Dmitriev, A. A. Vasilcov, I. S. Kostarev, J. Test. Diagn. **11**, 221 (2012)
4. W. Fan, P. Qiao, Structural Health Monitoring **10 (1)**, 83 (2010)
5. X. Guan, G-T. Zheng, Q.-W. Wang, JOA **7**, 1870 (2010)
6. M. Wagner, S. Airey, G. Piret, P. Le, *European space agency – ESA new reaction wheel characterization test facility (RCF)*, 142 (2012)
7. M. Pastor, M. Binda, T. Harcarik, Procedia Engineering **48**, 543 (2012)
8. D. J. Ewins, Modal Testing – theory, practice and application **5**, 60 (2013)
9. Y. T. Leung, JOSV **149**, 83 (1991)
10. Yu. A. Britova, A. V. Plotnikov, *Materialy nauchno-tehnicheskoj konferencii molodyh specialistov «Informacionnye sputnikovye sistemy» imeni akademika M.F. Reshetneva* **1**, 213 (2011)

MATEC Web of Conferences 48, 01006 (2016)
DOI: 10.1051/matecconf/20164801006

Designing of space oriented body parts with the use of modern technologies

Munkhe-Zul Ayusheev[1,a] and Tamara Kostyuchenko[1]

[1] National Research Tomsk Polytechnic University,634050 Tomsk, Russia

Abstract. Proper selection of the gating system for investment pattern is a long and labor-intensive process that requires significant resources. The latest design technologies enable the calculation of the gating system and simulation of the casting process based on these calculations. Simulation and calculation of the gating system make it possible to determine defects which occur during the casting process at the design stage. The result of simulation is the distribution of flow and temperature velocity vectors. The process of cooling and crystallization is simulated, as well as a mold filling process. Analyzing these data and changing the characteristics of the gating system, it is possible to achieve the very high quality of castings.

1 Introduction

The technology of investment casting is frequently used in the manufacture of devices, parts of space-aimed devices and particularly of body parts. Design of a cast part is technological if itis possible to produce the casting that meets the requirements of accuracy, surface roughness, physical and mechanical properties and structure of metal with the lowest production costs. The dimensions of the castings produced by the investment casting are maximally close to the dimensions of the completed part. Therefore, the cost of the product is shortened by the reduction of machining.

Foundry is one of the most highly-productive, fast, universal and cost-effective methods of metal ware molding. It was always important to create such technological processes which would ensure high operational life of parts by production of high-quality castings with the given structure and characteristics.

2 Modeling

The design of the gating system consisting of feeders, riser vents, downspures, sprue wells and bases plays an important role and has a direct impact on the quality characteristics of the casting. The quality of the casting process is directly related to the quality of an investment casting. The process of casting inevitably entails the defects of intimate structure of castings (cracks, sponginess, flux contamination and incomplete filling-up).

The process of casting and its quality depend on both the design of the gating system and its right selection. Construction of the gating system is currently based solely on the experience of the design

[a] Corresponding author : muncko94@mail.ru

engineer; therefore, there is a need in modeling, optimization and automation of a process of casting design using the CAD-systems to ensure the quality of casting. Simulation of the processes of casting, solidification and filling-up with the use of the latest technologies has become a necessity in the foundry industry [1].

Simulation of a casting process with the latest technologies makes it possible to eliminate possible casting defects in the early stages of the design which is a huge breakthrough in the field of mechanical engineering.

Thus, the modeling and design of the casting process using modern specialized computer programs help to reduce the production costs and increase the production and quality of the casting several times.

The crystal structure of the casting and the main processing, qualitative and operational characteristics depend on the number, shape, crystal growth rate and their initial orientation in the mold. Therefore, the important task in the field of foundry industry is the development of a methodology that ensures the monitoring of the structure and its characteristics, the heat and mass transfer, hydrodynamic casting processes and thermal-deformation interaction of casting and mold. The development of revolutionary new low-waste and non-waste technologies which provide a significant increase of operational, physical and mechanical characteristics of the cast metal is important as well [2]. New technologies will help to solve these problems.

The powerful software is designed in order to improve the efficiency and provide a high accuracy of numeric calculation. Every year, million tons of castings are produced worldwide, so the numeric calculations using the latest technology with the appropriate software are very necessary for the production of solid and large-sized castings [3].

Modern programs can read the STL file format of a solid geometric casting pattern. That is well since all CAD-systems work with the STL format and can fix the weakness of a STL file automatically without human intervention. Post-treatment may indicate the filling of a sample, 3D velocity change vector, 3D temperature distribution at any injection area and these operations are simple and easy [4]. Programs are able to select the different kinds of materials in order to define the frame properties during the numeric calculations of mold filling and solidification of the casting which allows the practical application of engineering analysis.

3 Methodology

The process of modeling and design of the gating system begins with the creation of a solid model, a functional surface model, drawings, technical specifications. New modern programs define the functional surfaces; dimensioning specifications, construction tolerances and boundary conditions using certain modules and based on these data, they model a parametric model wherein the depths of surfaces and geometrical dimensions are parametric. Engineer reuses parametric abilities of the gating systems models for the optimization of casting properties. The models are used to study the thermal, mechanical and structural characteristics.

That is a simulation model which will be a product model for various tests, particularly functional test, and based on these tests results the final model of the gating system is formed [5]. During the creation of a casting model the castability is changed by changing the geometry until the indicators of shrinkage, hole and recess filling meet the requirements.

4 Main characteristics

The main objective of the simulation of casting processes is the calculation of heat distribution in the investment casting with increase of time. This operation is carried out in specialized units. The fundamental heat transfer equation is determined by the thermal conductivity law. It is a specific differential equation of temperature (T) and time (t) relationship in the Cartesian coordinates x, y, z, where ρ is a density, Cp is a thermal capacity and λ - thermal conductivity [3]:

$$pc_p \frac{\partial T}{\partial t} = \frac{\partial}{\partial x}\left(\lambda \frac{\partial T}{\partial x}\right) + \frac{\partial}{\partial y}\left(\lambda \frac{\partial T}{\partial y}\right) + \frac{\partial}{\partial z}\left(\lambda \frac{\partial T}{\partial z}\right) + \dot{Q} \quad (1)$$

Therefore, the time derivative of temperature at a point is determined by the flow of heat quantity and the source-heat dissipation term. It is possible to solve one system of linear equations for one time step using the discrete partial equations and final element analysis with boundary conditions.

The gap appears between the casting and the casting mold during solidification and liquid-metal cooling which affects their heat transfer and solidification conditions. In each case the gap differs depending on the molding method, the heating rate of the mold and the nature of its interaction with the metal induration.

The gap formation and modification is determined by three factors: the shrinkage of the casting, mold deformation during the molding cycle, the moldable extension of the under the action of ferrostatic pressure. The main contribution to the formation of the gap and the kinetics of its change is made by the first two factors. The calculations are made with the use of software based on different methods of calculation [2].

The temperature field of the casting mold varies with the change of time as well as the temperature field of the casting during the cycle of induration. The gap setting is the sum of the absolute values of the casting's shrinkage and deformation of the casting mold. The temperature distribution during filling of the form for one of the variants of the gating system design is shown in Figure 1.

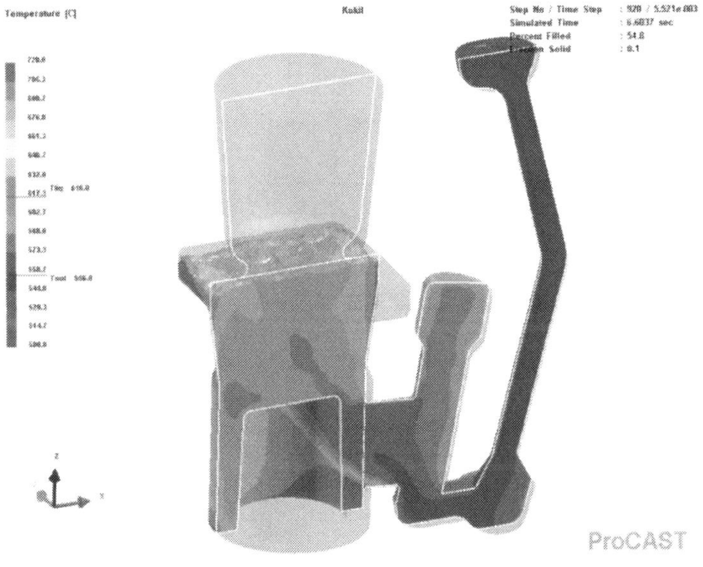

Figure 1. Filling the mold with the distribution of metal temperature. It is possible to define the temperature of liquid metal from the color using the grid.

The modern programs of computer modeling of foundry technologies based on physical theories of the thermal, diffusive, hydrodynamic and deformation phenomena are capable to imitate adequately many processes happening when filling with fusion of a form, at crystallization of a multicomponent alloy and further cooling of casting.

5 Conclusion

Computer simulation of casting processes and solidification has become a necessity in the foundry industry. The parameterization makes it possible to optimize quickly the casting properties of the casting which meet all the requirements within a very short time. Computer programs can read the STL file, a complex solid geometry, show the velocity vector, 3D velocity and temperature fields, the filling of a sample, distribution of mold filling temperature and solidification.

It is possible to calculate the velocity and temperature fields for the laminar and turbulent flows during the filling of the mold and induration process. Numeric simulation can be used in order to predict the casting defects and improve the characteristics of the casting process for creation of body parts for space oriented equipment.

References

1. A. Swapnil, S. B. Ambekar, Dr. Jaju, Int. J. Eng. Res. and Gen. Scien. **2**, 93 (2014)
2. E. I. Marukovich, V. F.Bevza, A. M.Bodyako, U. P.Hrusha, Cast. and Metal. **3**, 63 (2011)
3. W. Junqing, S. Xun, G. Yang, W. Penghua, L. Hailan, B. Limei, S. Xinzhi, China Fdy., 179 (2008)
4. P. Malinowski, J. S. Suchy, J. Jakubsky, Arch. of metal. and ma. **3**, 965 (2013)
5. C. Clauss, S. Schuch, R. Finocchiaro, S. Lankes, T. Bemmer, *IPDPS*, 306 (2006)
6. J. Jokinen, P. Makkonen, T. Saarelainen, E. Coatanéa, *ICED* **7**, 28 (2007)

MATEC Web of Conferences 48, 01007 (2016)
DOI: 10.1051/matecconf/20164801007

Power processing unit for hall-effect thrusters on "Meteor-M №3 spacecraft"

Anton Moshnyakov[1, 2, a], Maxim Mikhaylov[1, 2]

[1] National Research Tomsk Polytechnic University, 634050 Tomsk, Russia

Abstract. The development results of power processing unit (PPU-M) for hall-effect thrusters on «Meteor-M №3» spacecraft are considered. The structure, weight, dimensions and main technical characteristics of the system in the paper are presented. The work peculiarity of the system is unstable input voltage of both power bus and control bus that increases the ripple voltages and currents at the input and the output and causes the additional requirements to the circuit design. A comparative analysis of the system characteristics and European analogs was carried out, and then a conclusion on the basis of available data that the characteristics of the system are not inferior to European analogs was made.

1 Introduction

Power processing unit provides the required modes of two SPT-100 (Stationary Plasma Thruster) thrusters installed on board of «Meteor-M №3» spacecraft and designed to station-keeping during the whole lifetime (7 years) and then to inject into Graveyard orbit. PPU-M is suitable for operation in open space and provides the required power supply modes for the loads shown in Table 1.

Table 1. Loads of PPU-M.

Load type	SPT-100	Thruster valves	Xenon storage bank valves	Pyrovalves	Xenon flow regulators
Quantity, pcs	2	12	6	2	2

2 PPU-M Design

PPU-M is designed as a monoblock and has two channels of power supply, each of which supplies the power for one thruster (Figure 1).

Structurally, PPU-M consists of 14 modules installed on a common base, connected by a bus and closed by a protective cover. Figure 2 shows a power supply circuit of one SPT-100 thruster type.

[a] Corresponding author : polus@online.tomsk.net

[2] JSC «Scientific and industrial Centre «Polyus», 634050 Tomsk, Russia

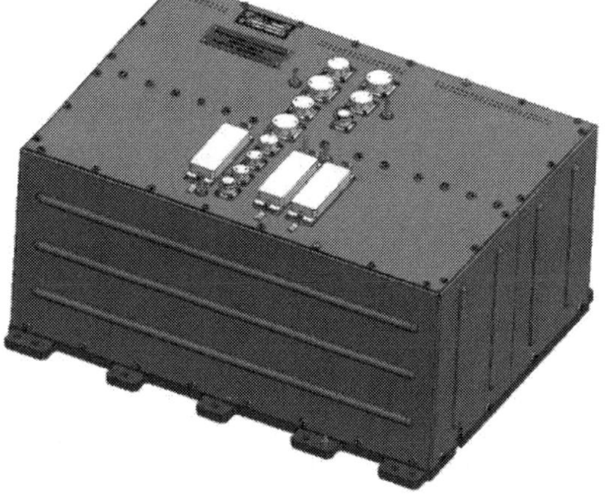

Figure 1. PPU-M mechanical design.

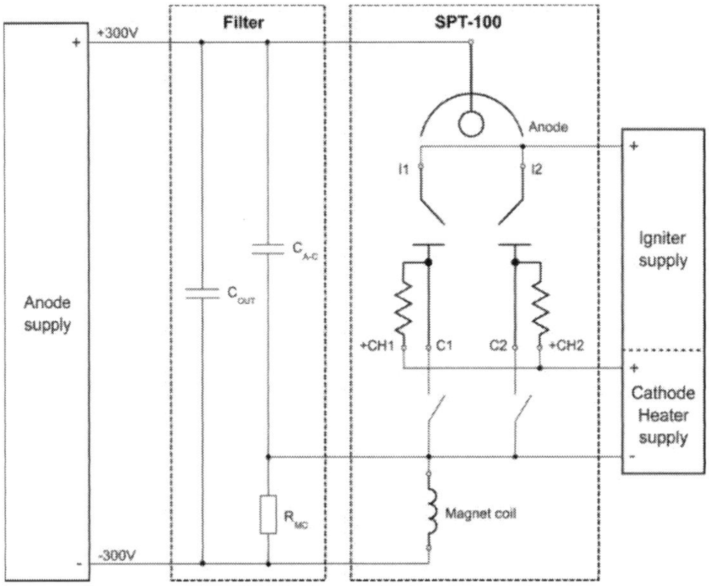

Figure 2. Power supply circuit of SPT-100: CH – Cathode Heater; C – Cathode; I – Ignition Electrode; $C_{A\text{-}C}$ – anode-cathode capacitor; C_{OUT} – output capacitor; R_{MC} – magnetic coils ballast resistor.

3 PPU-M Specifications

PPU-M work peculiarity is unstable input voltage: in the power bus it varies in the range of 46 - 65 V, and the control bus in the range of 24 - 34 V. Pulsed power consumption from input sources and as a result increasing of their ripple currents and voltages are caused by this fact. A throttle is installed and value of anode supply input filter capacitor is increased to reduce them. The value of C_{OUT} capacitor is increased to reduce the voltage ripple in the anode circuit. The value of $C_{A\text{-}K}$ capacitor will be determined in the process of thruster connecting tests.

The power supply circuit of magnetic coils for SPT-100 is worked out using discharge current that eliminates additional source and reduces the weight of the device. For the same purpose the original cathodes circuit switching scheme is used, that allows switching cathode heater, cathode and the ignition electrode circuit using only one key.

In addition to the power supply sources, PPU-M has a built-in filter for protection against anode, cathode and ignition electrode circuit static electricity.

Main technical characteristics of PPU-M are given in Table 2.

Currently laboratory PPU-M sample is manufactured and developmental testing is being completed.

Table 2. Main characteristics of PPU-M.

Parameter	**Value**	**Note**
Discharge voltage, V	300±15	DC
Discharge current, A	4,5±0,1	In a start mode up to 7 A for 2 - 4 sec
Cathode heater current, A	12,0±0,5	0,2 Ohm < R_{CH} < 0,7 Ohm
Xenon flow regulators current, A	1,5 – 1,75	Preparation mode
	0; 3,8±0,2	Operating mode
Ignition pulses voltage, V	290 – 350	R_{IE} = 10±1 kOhm; f = 145±15 Hz; t_I = 70 – 140 mks
	20 – 40	R_{IE} = 5,0±0,5 Ohm; f = 145±15 Hz; t_I = 70 – 140 mks
Thruster valves voltage, V	22,3 – 32,7	Switch-on mode
	8 – 13	Hold mode
Xenon storage bank valves voltage, V	22,3 – 32,7	Switch-on mode
	8 – 13	Hold mode
Pyrovalves current, A	3 – 5	0,8 Ohm < R_{PV} < 1,2 Ohm
	0,005 – 0,015	Flow mode
Power bus voltage, V	46,9 – 65,3	Power consumption in operating mode less than 1500 W
Control bus voltage, V	24 – 34	Power consumption less than 80 W
Command quantity, pcs	17	Single commands
The number of telemetry channels, pcs	8	Analog, 0 V < U_{TM} < 5 V
	8	Signal
Reliability	0,999	Active lifetime is 7 years long; operating time is 1000 hours long
Weight, kg	17	

4 Comparative analyses

Existing analogues characteristics analysis was performed in the development phase. Main results are shown in Table 3.

The nearby analogue of the PPU-M is PPU Mk1 for Stentor spacecraft [1] because both of them operate with two thrusters type SPT-100. A significant difference in the weight of these PPUs is due to the principally different structure of design. PPU Mk1 for Stentor spacecraft contains one power supply channel and one switch module for two thrusters [2], a single failure of any element results in failure of the whole system. Only «cold» redundancy using two PPU on board is operated in this spacecraft.

In addition, unlike the PPU for Stentor and Smart-1 spacecraft, powered by a stabilized [3] power bus 50 V, in PPU-M the power supply is unstabilized (24 –34 and 46 – 65) that includes a built-in PPU stabilizer. Also the PPU-M includes a built-in filter for protect against electrostatic discharges, which is submitted as a separate and independent unit of the propulsion system in European PPU.

Table 3. Main specifications of several domestic and foreign PPU-M for thrusters SPT-100.

Name of system	Start of operation	Thruster		Weight of PPU, kg	Power supply bus, W	Anode supply		Cathode heater current I_{CH}, A
		Type	Quantity			U_A, V	P_A, kW	
PPU - 2EA (SIC "Polyus")	2013	SPT-100	8	27,1	27, 100	300	1,4	12 (DC)
PPU-M (SIC "Polyus")	2015	SPT-100	2	17	24 – 34; 46,9 – 65,3	300	1,4	12 (DC)
PPU Mk1 for Stentor spacecraft	2002	SPT-100 or PPS-1350	2	10,4	100 or 50	300 or 350	1,6	14 (DC)
PPU Mk1 for Smart-1 spacecraft	2003	PPS1350-G	1	10,9	50	175 – 350	0,306 –1,225	No data

Based on the above it can be concluded that the build structures of PPU-M and PPU (Stentor) are significantly different, that greatly complicates the identification of their specific characteristics. To ensure a high level of reliability for PPU it is required to install a redundant unit. The vast majority of foreign PPU-M analogues have a remote filter unit, whose weight is not considered.

5 Conclusion

Designed PPU-M ensures reliable operation of two SPT-100 type thrusters as part of «Meteor-M №3» spacecraft in open space with unstable input voltage. For functional and specific weight and size characteristics PPU-M is superior to foreign analogs due to the original circuit switching cathodes and monoblock design.

References

1. E. Bourguignon, T. Scalais, J. Thomas, ESA SP **555**, 447 (2004)
2. P. Garnero, T. Grassin, F. Darnon, L. Petitjean, IEPC, 048 (2003)
3. E. Bourguignon, S. Fraselle, T. Scalais, IEPC, 095 (2011)

MATEC Web of Conferences 48, 02001 (2016)
DOI: 10.1051/matecconf/20164802001

Stiffness estimation for large–sized umbrella space reflector

Alexey Belkov[1], Sergey Belov[1,a], Mikhail Pavlov[1], Victor Ponomarev[1], Sergey Ponomarev[1] and Andrey Zhukov[1]

[1] *National Research Tomsk State University, 634050 Tomsk, Russia*

Abstract. This work describes stiffness estimation for new large-sized (with aperture more than 50 m) umbrella space reflector. Reflector stiffness is measured by natural frequencies. Mode shapes and respectively natural frequencies are calculated by finite element method. First four natural frequencies with respectively mode shapes are presented.

1 Introduction

Currently, large-sized deployable space reflectors (antennas) are widely used due to their effective deployment within significant coverage per relatively small mass. Such antennas are usually classified as hoop («AstroMesh») and umbrella («Halca», «Garuda-1») type [1, 2].

One of the major requirements for the space reflector frame is stiffness level i.e. the mechanical design property to resist external factors. However, there exists inconsistency between required high stiffness and increasing geometrical coverage of space reflectors to space payload weight, which in its turn, would influence the missile vehicle launching in to space orbit. To retain weight limitation under conditions of increasing reflector sizes, it is essential to decrease cross section area of frame elements which results in lower frame stiffness. In this case numerical stiffness analysis is an important phase in designing these systems.

2 Natural frequency spectra characterizing stiffness of mechanical systems

As a result of external loads the mechanical systems deforms - the more the stiffness, the less the deformation.

Large-sized space reflector is tensioned cable-stayed structures with thin tensile metallic mesh, forming the reflective surface. Such mechanical systems should satisfy the requirements affecting the mechanical behavior itself. In particular, reflector should retain its optimal reflector surface shape under specific operation loads, while root mean square (RMS) error of reflecting surface should not be more than 2-3% of the operating wavelength.

To estimate space reflector stiffness level, the spectrum of natural frequencies is applied. Natural frequency values in estimating stiffness can be shown in terms of one - dimensional mechanical system with focused parameters, which, in its turn, demonstrates finite stiffness [3].

[a] Corresponding author : belovsv@niipmm.tsu.ru

In [3] expression for mechanical system amplitude is presented as a function $A = A(f, m, \omega, \omega_0)$:

$$A = \frac{f}{m} \cdot \frac{1}{\sqrt{(\omega_0^2 - \omega^2)^2}}, \tag{1}$$

where, f – external force amplitude, m – mass, ω – external force frequency, ω_0 – natural frequency.

The most preferable case is $\omega_0 >> \omega$. In this case the mechanical system relative to external dynamic load would behave as a rigid body. In the above formula it is important that dynamic stiffness depends on natural frequency and external force amplitude.

The most negative case is when $\omega_0 = \omega$, where the mechanical system experiences a resonance. This resonance causes $A \to \infty$. For real systems amplitude is damped either by properties of materials, joint defects and special damper or they are destroyed.

In static case, when $\omega \to 0$,

$$A = \frac{f}{m} \cdot \frac{1}{\omega_0^2}, \tag{2}$$

there is a static deflection. In this case this value indicates stiffness and has the following dimension N/m. Thus, the parameter ω_0 characterizes stiffness of the mechanical system – the higher the natural frequency, the less the static deflection.

As the beam elements, forming the framework, are of significant length, beam stiffness is either comparable to or less stiff than the joints. Consequently, the reflector is a mechanical system with parameter distribution. To determinate the stiffness of such structure, finite element method is applied to calculate natural frequencies, as the analytical approach is entirely unsuited for this. This can be explained by the fact analytical approach is suited for problems with small dimension.

To define the natural frequencies of the reflector it is necessary to know the stress and strain state of the reflector itself. Thus, determining the natural frequencies of the reflector involves two stages:

- stress and strain state definition for min. RMS error for reflector surface;
- natural frequencies definition at stress and strain state of the reflector.

The first stage is described in [4], therefore, only the second stage is described in details.

3 Modal analysis

In finite element discretization, motion equations for elementary dampingless volume with free oscillations are:

$$\mathbf{M}\ddot{\mathbf{u}} + \mathbf{K}\mathbf{u} = \mathbf{0}, \tag{3}$$

where $\mathbf{M}$ – mass matrix, $\mathbf{u}$ – displacement vector for each element node, $\mathbf{K}$ – stiffness matrix for all structure elements.

Equation (3) solution is:

$$\mathbf{u} = \boldsymbol{\varphi} \cdot cos\omega t, \tag{4}$$

where, $\boldsymbol{\varphi}$ – vector of mode shapes, ω – circular frequency.

Subtituting (4) into (3), we obtain the equation (5) :

$$\left(\mathbf{K} - \omega^2\mathbf{M}\right)\boldsymbol{\varphi} = 0. \tag{5}$$

The circular frequencies ω are obtained by the equation (6):

$$\left|\mathbf{K} - \omega^2 \mathbf{M}\right| = 0. \qquad (6)$$

Vectors $\boldsymbol{\varphi}_i$ which correspond to circular frequencies ω_i are obtained by equation (7):

$$\left(\mathbf{K} - \omega_i^2 \mathbf{M}\right)\boldsymbol{\varphi}_i = 0, \qquad (7)$$

where $i = \overline{1, N_{DOF}}$, N_{DOF} - number degree of freedoms.

This is eigenvalue problem. It is solved by Block Lanczos method [5]. Natural frequencies can be expressd by circular frequences:

$$f_i = \omega_i / 2\pi. \qquad (8)$$

Expression (8) we use for stiffness estimation of umbrella reflector.

4 Umbrella reflector description

Figure 1 shows finite element models of the umbrella type reflector in sections and assembled section.

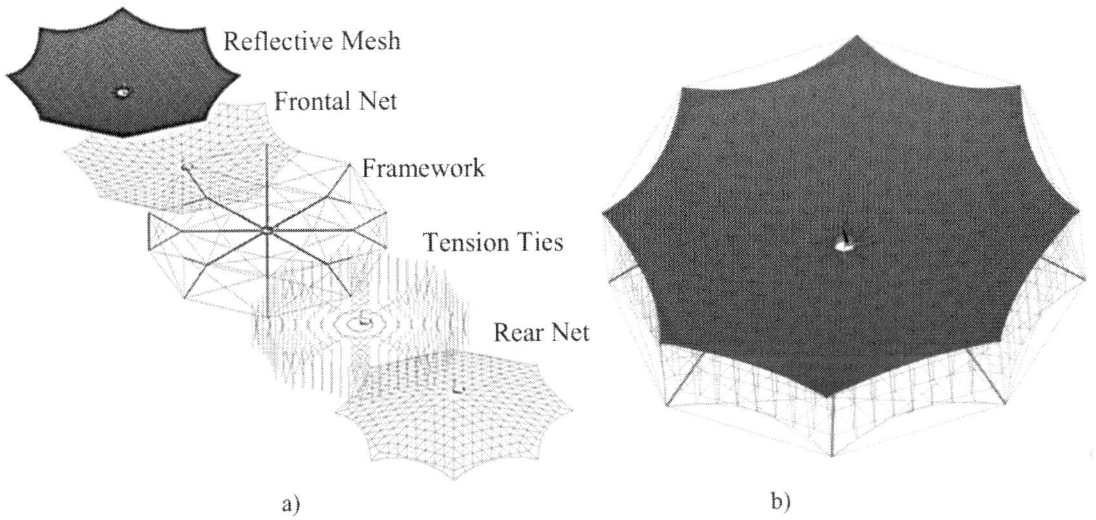

Figure 1. Umbrella reflector: a) Reflector sections; b) Assembled reflector.

One can see that the reflector consists of reflective mesh which is modeled as a membrane without flexural stiffness, frontal net, rear net, tension ties which are modeled as cords with constant cross section area, which are subjected to tension. Framework is modeled of telescopic spokes which consist of grid-type beam elements the length of which can be seen in Figure 2. All spokes are gathered at central hub which is modeled by shell and beam elements.

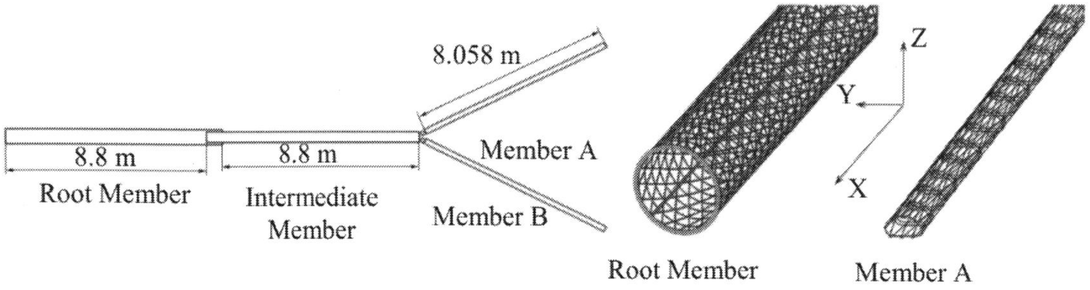

Figure 2. Grid-type telescopic spokes.

Diameters of the root and intermediate members are of 0.308 m and 0.256 m, respectively. For A and B members are max of 0.18 m and min 0.095 m diameters.

4.1 Reflective surface geometry description

Reflective surface consists of frontal net and reflective mesh. It's geometry (so-called offset paraboloid) is the intersection of revolution paraboloid with focal distance – 32 m and cylinder of diameter 40 m. Cylinder and paraboloid axes are parallel and separated by a clearance of 8 m (figure 3 a). Figure 3 b) shows topology of the frontal net.

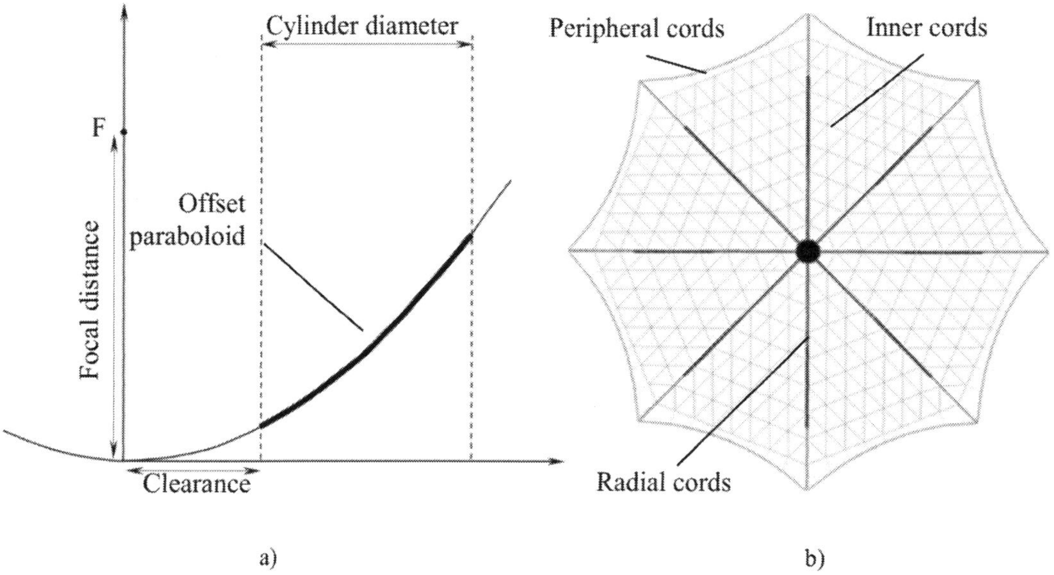

Figure 3. Reflecting surface geometry: a) Offset reflector parameters b) Frontal net topology.

Plane geometry of peripheral cords are defined by the parabola in the equation $y(x)=\left(4D/L^2\right)x^2-D$, where L – distance between ends of A members of neighboring spokes, $D>0$ defines sag of the parabola.

Inner cords have diameter of 0.7 mm, tension ties of 0.4 mm. All other cords have a diameter of 2 mm. Reflective mesh has a thickness of 0.1 mm.

4.2 Mechanical characteristics and properties of materials

Table 1 defines mechanical characteristics of spokes and their joints. It includes bending stiffness of all spoke members, member joint torsion stiffness. Table 2 defines properties of cord materials.

Table 1. Mechanical characteristics.

Member	**Bending stiffness, N·m^2**	
Root Members	$1.2620 \cdot 10^6$	
Intermediate Members	$7.6937 \cdot 10^5$	
A and B Members	Max Diameter $2.8708 \cdot 10^5$	Min Diameter $5.2931 \cdot 10^4$
Joints	**Torsional stiffness on *X* axis, N·m/rad**	**Torsional stiffness on *Y*, *Z* axes, N·m/rad**
Root Member and Hub Joints	$4.4 \cdot 10^5$	$7.5 \cdot 10^5$
Root and Intermediate Member Joints	$2.0 \cdot 10^5$	$1.00 \cdot 10^6$
Intermediate and A (B) member Joints	$6.67 \cdot 10^4$	$1.33 \cdot 10^5$

Table 2. Properties of cord materials.

Cords	**Module of elasticity, Pa**	**Density kg/m^3**
Peripheral, Radial Cords	$9.83 \cdot 10^9$	2450
Inner Cords	$3.27 \cdot 10^{10}$	2339
Cords Joining of A and B Members Ends	$9.83 \cdot 10^9$	8000
Reflective Mesh	$1.3 \cdot 10^6$	0.02 kg/m^2

5 Numerical results for natural frequencies and mode shapes

Natural frequencies and, respectively, mode shapes are calculated using the finite element method. Boundary conditions are: pretentions in inner cords – 10 N, pretentions in peripheral cords – 250 N, pretentions in tension ties – 1 N, pretentions in reflective mesh – 2 N/m, pretentions in cords which connect ends of A and B members – 90 N, pretentions in others cords – 100 N. Reflector is fixed at central hub. Figures 4-5 show mode shapes and, respectively, natural frequencies. Arrows on figures 9-10 show oscillation directions.

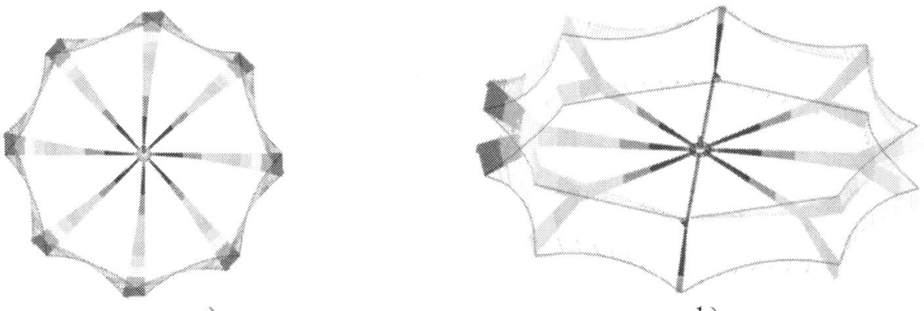

a) b)

Figure 4. First and second mode shapes with frequency of: a) 0.525Hz; b) 0.546Hz.

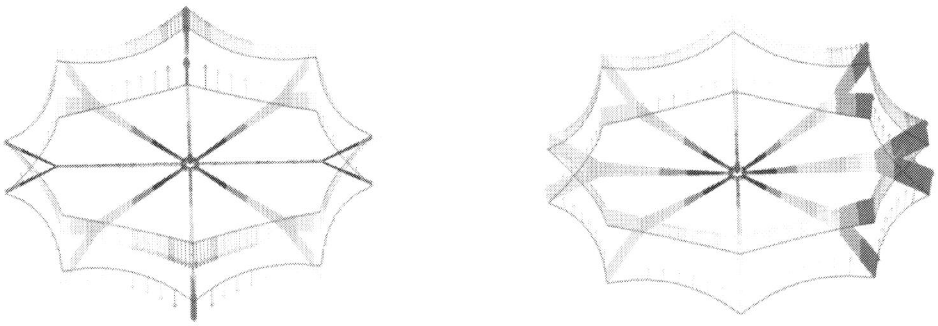

a) b)

Figure 5. Third and fourth mode shapes with frequency of: a) 0.577 Hz; b) 0.647 Hz.

6 Conclusion

The present work describes the stiffness estimation based on natural frequencies for new large sized umbrella type space reflector. A link between natural frequencies and static deflection is identified. A finite element model is designed for described reflector. As a result of numerical analysis mode shapes and, respectively, frequencies are presented.

Acknowledgement

This work has been financially supported by The Ministry of Education and Science of The Russian Federation. Unique identifier RFMEFI57814X0073.

References

1. S. Morterolle, B. Maurin, J. Quirant, C. Dupuy, AA **76 (1)**, 154 (2012)
2. A. G. Tibert, *Thesis work* (Department of Mechanics, Royal Institute of Technology, Sweden, 2002)
3. S. Timoshenko in collaboration with D. H. Young, *Vibrations Problems in Engeneering*, *Third edition* (D. Van Nostrand Company, Inc, Toronto- New-York - London, 1955)
4. S. Ponomarev, A. Zhukov, A. Belkov, V. Ponomarev, S. Belov, M. Pavlov, IOP Conf. Series: M SE **71** 012070 doi:10.1088/1757-899X/71/1/012070 (2015)
5. K.-J. Bathe, *Finite element procedures* (Prentice-Hall. Englewood Cliffs, 1996)

MATEC Web of Conferences **48**, 02002 (2016)
DOI: 10.1051/matecconf/20164802002

Sensitivity improvement techniques for micromechanical vibrating accelerometers

Sergey Vtorushin[1, a] and Tamara Nesterenko[1]

[1] *National Research Tomsk Polytechnic University, 634050 Tomsk, Russia*

Abstract. The paper presents the problems of detecting a desired signal generated by micromechanical vibrating accelerometer. Three detection methods, namely frequency, amplitude and phase are considered in this paper. These methods are used in micromechanical vibrating accelerometers that incorporate a force sensitive element which transforms measured acceleration into the output signal. Investigations are carried out using the ANSYS finite element program and MATLAB/Simulink support package. Investigation results include the comparative analysis of the output signal characteristics obtained by the different detection methods.

1 Introduction

Micromechanical vibrating accelerometers (MMVA) have been widely spread rather recently. Much attention is paid to these accelerometers due to their specific properties and operational characteristics that essentially differ from that of micromechanical capacitive accelerometers (MMCA) used today. The MMVA advantages include the high sensitivity that exceeds that of traditional capacitive accelerometers by one order more; high dynamic range and stability [1].

Presently, the microelectromechanical systems (MEMS) are manufactured by leading companies of the USA, Japan, Korea, China, and other countries. The Draper Laboratory is an American not-for-profit research and development organization that one of the first started the investigations of silicon MMCA. Accelerometers designed and produced by the Draper Laboratory possess the stability of scaling factor of not less than 1 ppm [2]. MMVA designed in the University of California, Berkeley, possesses the sensitivity of 17 Hz/g. This accelerometer incorporates two oscillators that provide the decrease of its thermal sensitivity [3]. Researchers from Seoul National University (South Korea) designed the micromechanical resonator silicon accelerometer having 70 Hz/g sensitivity at 12 kHz resonance frequency of oscillators [4]. Chinese MEMS developers are currently involved in modeling silicon accelerators. Thus, for example, Tsinghua University (Peking) in cooperation with the Institute of Microelectronics developed the MMVA prototype having 27,3Hz/g sensitivity and 167,8 μg resolution [5]. The past decade has seen remarkable advances in the development of MEMS technologies owing to their high reliability and low costs [6].

[a] Corresponding author : vtorushin1991@mail.ru

2 Operation principle of single-axis MMVA

The principle of operation of any micromechanical vibrating accelerometer incorporating a sensitive element (oscillatory-type transducer), is based on the difference between the transducer eigenfrequencies induced by its stess-strain state. A force sensitive element or silicon resonator shown in Figure 1 can be used as an oscillatory-type transducer.

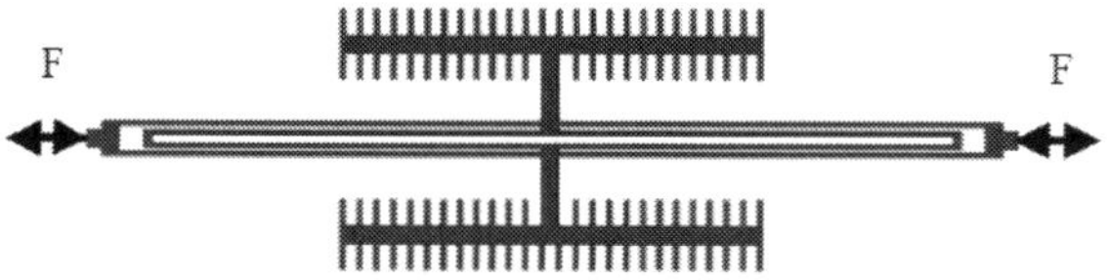

Figure 1. Force sensitive element.

Under forces F, the resonator eigenfrequency is changed depending on the force direction. The resonator eigenfrequency can vary not only under longitudinal and transverse forces but also force moments. At the same time, the maximum eigenfrequency unbalance is observed under the longitudinal forces [7].

The change in resonator eigenfrequency induced either tensile or compressive forces can be calculated from

$$\Delta f = i^2\pi^2\sqrt{\frac{EI_x}{ml^4}}\sqrt{1\pm\frac{\Delta Fl^2}{i^2\pi^2 EI_x}} \tag{1}$$

In (1), the first three multipliers describe the eigenfrequency mode i without considering the pre-stressed state of resonator. The last multiplier describes the longitudinal force F the positive sign of which corresponds to tensile force, while the negative – to compressive. The changes in eigenfrequency and longitudinal force are proportional to each other.

This principle can be used to design a sensitive single axis vibratory accelerometer shown in Figure 2.

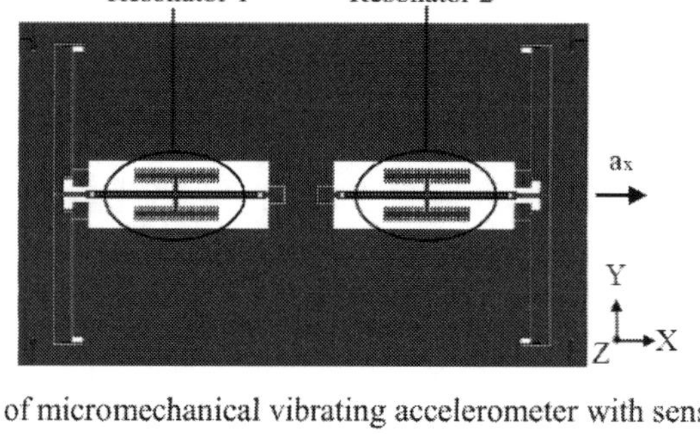

Figure 2. Schematic structure of micromechanical vibrating accelerometer with sensitive single axis.

The vibratory micromechanical accelerometer is provided with the inertial mass connected with the frame by the elastic wires and by the lever system with resonator. Under the external acceleration ax, the inertial mass is subjected to inertial forces transmitted by the lever system to silicon resonators. The lever system provides several times increase of the transmitted force depending on the arm. Internal stresses occurred in resonators produce changes in their eigenfrequencies, resonator 2 being extended and resonator 1 being compressed. The eigenfrequency shifting reflects the amount of the actual acceleration.

Thus, the lever system improves the sensitivity of accelerator to the acceleration measured, while the differential resonator connection circuit allows constructing the two-channel acceleration transforming circuit with the common-mode algorithmic compensation of thermal noise.

3 The model of single-axis MMVA

The ANSYS domain modal analysis allows obtaining two working vibration modes of the accelerometer (Figure 3).

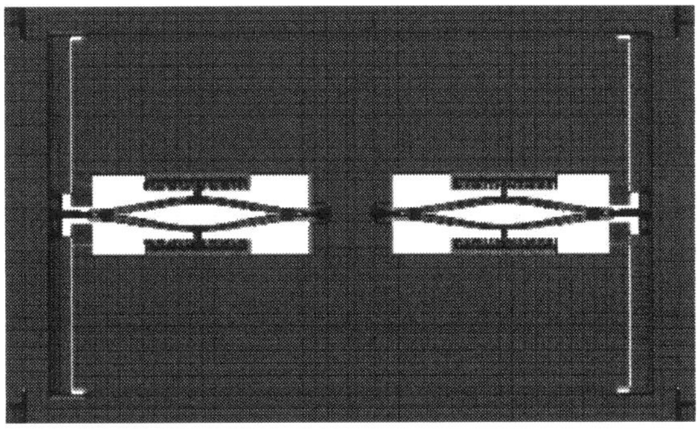

Figure 3. Schematic structure of micromechanical vibrating accelerometer with the fourth frequency mode.

Vibration excitations of silicon resonators are observed at the fourth frequency mode. Each couple of strings of the both resonators is excited in antiphase, thereby compensating the forces and force moments generated by each string and decreasing the energy out feed in the environment through the supports.

The two-channel differential resonator connection circuit increases twice the accelerometer sensitivity. According to (1), the eigenfrequency of resonator subjected to tensile force increases proportionally to the force F. The eigenfrequency of the second resonator decreases due to the compressive force. The total change in eigenfrquiency is calculated using the following equation:

$$\Delta f_{\Sigma} = i^2\pi^2\sqrt{\frac{EI_x}{ml^4}}\left(\sqrt{1+\frac{\Delta Fl^2}{i^2\pi^2 EI_x}} - \sqrt{1-\frac{\Delta Fl^2}{i^2\pi^2 EI_x}}\right) \tag{2}$$

As a result, we obtain the doubled change of the eigenfrequency. According to (2), the accelerometer sensitivity to longitudinal forces and, as consequence, measured acceleration can be improved when decreasing the flexural stiffness EI_x of resonator strings and varying their length l.

In accordance with the ANSYS domain modal analysis accounting for pre-stressed state of resonator, the output characteristics are ascertained for the accelerometer. Figures 4 and 5 contain plots of these characteristics.

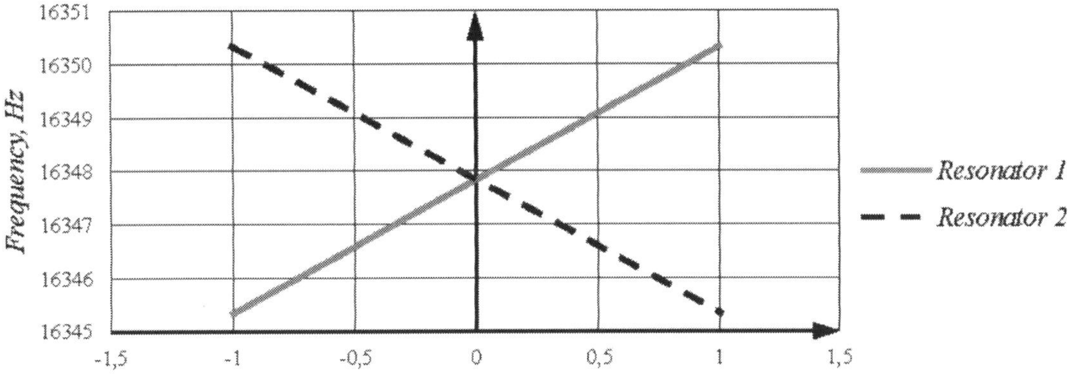

Figure 4. Output characteristic under X-axis acceleration.

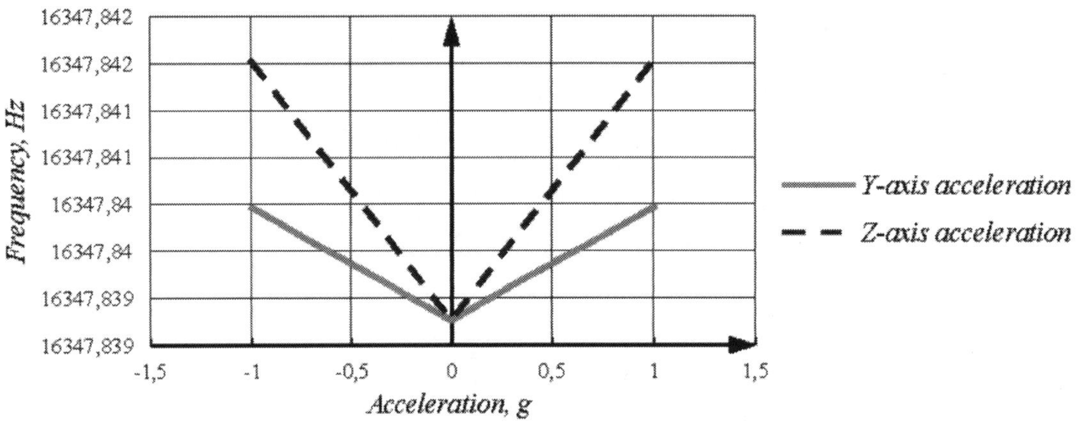

Figure 5. Output characteristic under *X*- and *Y*-axis acceleration.

Affected by the *X*- and *Y*-axis acceleration, eigenfrequencies are changed insignificantly only in one direction and can be completely compensated meeting the following conditions:

$$\Delta f_{\Sigma} = i^2\pi^2\sqrt{\frac{EI_x}{ml^4}}\left(\sqrt{1+\frac{\Delta Fl^2}{i^2\pi^2 EI_x}} - \sqrt{1+\frac{\Delta Fl^2}{i^2\pi^2 EI_x}}\right) \tag{3}$$

Herewith, the frequency difference under *X*-axis acceleration equals to 5,014 Hz in conformity with Figure 4 and equation (2).

3.1 Amplitude and phase detection technique

The amplitude and phase detection technique involves the measurement of the vibrational amplitude shift induced by acceleration. The maximum vibrational amplitude is observed at the coincidence of eigenfrequency of resonators and the excitation frequency of generator. The vibrational amplitude decreases at the eigenfrequency shift of resonators due to their stress-strain state. Depending on the mode of excitation, the operating point can be observed in the maximum peak of the amplitude-frequency curve or shifted both to left and right relative to this maximum. While selecting the mode of frequency mistuning with the operating point shift, it is possible to develop the differential measurement circuit.

The phase detection technique employs measurement of the phase difference of the reference-input signal produced by generator and vibrations produced by resonators. The phase difference at resonance is 90°. The eigenfrequency shift of resonators results in a change of the phase difference.

The mathematical simulation is used to calculate the accelerometer dynamics in order to evaluate each method by the maximum sensitivity criterion. The mathematical model presents the system of non homogeneous differential equations having the form

$$m_i\ddot{y}+\frac{\omega_i(q)m_i}{Q}\dot{y}+\omega_i^2(q)y=F_e sin(\omega t) \tag{4}$$

This equation is written in modal coordinates and complies with the typical equation of the second-order vibratory link. The transfer function for each string takes the form

$$W(s)=\frac{3.798\cdot 10^{17}}{s^2+1027s+1.055\cdot 10^{10}} \tag{5}$$

The amplitude- and phase-frequency curves of accelerometer are presented in Figure 6. The Q-factor is accepted to be 1000. The two curves describe the systems with (a = 1 g) and without acceleration (a = 0 g).

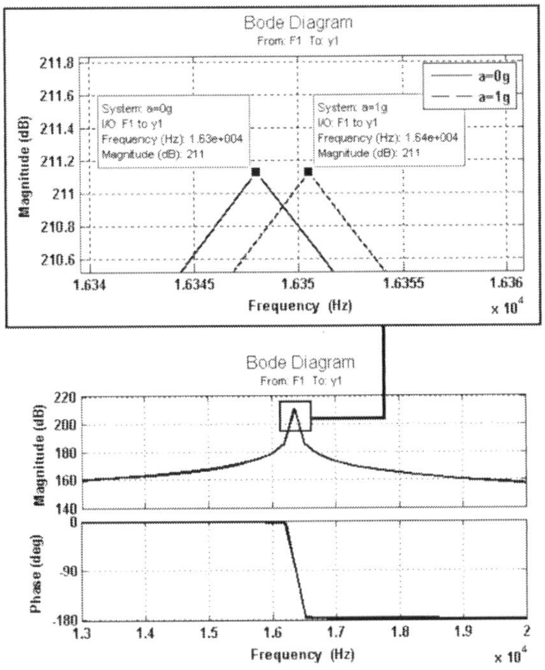

Figure 6. Amplitude- and phase-frequency curves for one string of resonator.

In Figure 7, the comparative characteristics of accelerometer are presented within the 0–1 g interval in accordance with the given frequency responses.

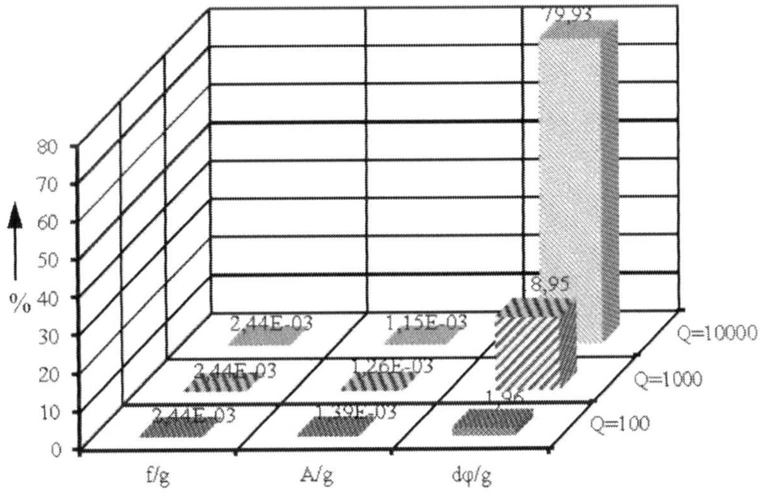

Figure 7. Comparative characteristics of accelerometer.

The maximum change is observed in the phase difference. The required frequency properties and $\Delta\varphi$ = 71,9 deg/g sensitivity can be obtained for the accelerometer at Q = 10000 achieved by the pressure decrease in the frame. The changes in frequency and amplitude under the acceleration less depends on the Q-factor and equal to ΔA = 0,00266 dB/g and Δf = 5,014 Hz/g, respectively.

These investigations showed that the maximum informativeness and sensitivity were achieved for the accelerometer with phase detection of a desired signal. The method of phase detection provides the design of accelerometer with the digital output, thereby considerably increasing its noise resistance.

4 Conclusion

These investigations showed that the maximum informativeness and sensitivity were achieved for the accelerometer with phase detection of a desired signal. The method of phase detection provides the design of accelerometer with the digital output, thereby considerably increasing its noise resistance.

Acknowledgement

This work was financially supported by the Ministry of Education and Science of the Russian Federation (agreement N 14.575.21.0068, unique identifier RFMEFI57514X0068).

References

1. D. W. Burns, R. D. Homing, W. R. Herb, J. D. Zook, H. Guckel, *Sens. Actuator* **53**, 249 (1996)
2. L. Huang, H. Yang, Y. Gao, L. Zhao, J. Liang, *Sensors* **13**, 15785 (2013)
3. A. A. Seshia, M. Palaniapan, T. A. Roessig, R. T. Howe, R. W. Gooch, T. R. Schimert, and S. Montague, Journal of microelectromechanical Systems **11**, 784 (2002)
4. H. C. Kim, S. Seok, I. Kim, S.-D. Choi, K. Chun, *Proceedings of the 13th International Conference on Solid-state Sensors, Actuators and Microsystems* **1**, 172 (2005)
5. Y. Jia, Y. Hao, R. Zhang, *Chin. J. Semicod* **2**, 281 (2005)
6. T. G. Nesterenko, S. E. Vtorushin, E. S. Barbin and A. N. Koleda, *IOP Conference Series: Materials Science and Engineering* **81** (2015)
7. S. E. Vtorushin, T. G. Nesterenko, Fundamental research **2**, 2553 (2015)

MATEC Web of Conferences 48, 03001 (2016)
DOI: 10.1051/matecconf/20164803001

Modeling axial 8-coil system for generating uniform magnetic field in COMSOL

Pavel Baranov[1,a] and Vitalia Baranova[1]

[1]National Research Tomsk Polytechnic University, 634050 Tomsk, Russia

Abstract. A changeable source of a uniform magnetic field is necessary for test and calibration of magnetometers. Often the Helmholtz coils is used for generating magnetic uniform field, but uniform field area generating by Helmholtz coils is confined to a small volume in the centre of the coils. The calculation and modelling results of an axial 8-coil system based on Helmholtz coils is described, which enables a magnetic field with a no uniformity of not more than 0.07 % to be produced at a distance of half the radius from the centre of the coil system.

1 Introduction

Measurement of magnetic induction as the characteristics of the magnetic field is one of the most important tasks in the field of creation and operation of navigation systems, orientation and stabilization. A standard of the DC magnetic field requires for test, calibration and graduation of sensors of magnetic fields and magnetometers based on them. As such a standard is applied system of one, two or three coils and a current source. A standard accuracy of this should be at least three times smaller than the error of the calibrated instrument.

The system of solenoid coils or Helmholtz coils [1-3] often used for generating uniform field and calibration magnetic fields sensors. General analysis of the magnetic field in Helmhotz coils and has been presented in works [4-8]. However, it is noticed that the uniform volume of magnetic field in the Helmhotz coil is not large enough for some applications. The calculation of an axial 8-coil system based on Helmholtz coils was described in paper [9].

In this paper we modeled 8-coil system based on Helmholtz coils in COMSOL and compared results.

2 Mathematical description

The Helmholtz coils (Figure 1) consist of two concentric rings of radius R arranged at a distance R from each other as shown in figure 1. The coils are connected in series with the current source to create a uniform field. According to the law of Biot-Savart-Laplace, the resulting field of the two coils is equal to the vector sum of the fields generating by each coil. The axial field of one coil of a predetermined point can be calculated according to the formula (1):

[a] Corresponding author : bpf@tpu.ru

$$B(z) = \frac{\mu_0 NIR^2}{2(R^2 + (z-h)^2)^{2/3}}, \tag{1}$$

where μ_0 – magnetic permeability of vacuum, H/m;
N – the number of turns of the each coil;
I – current through the coils, A;
z – coordinate of the point, m;
h – the distance of the coil center from the beginning of the coil center coordinates, m.

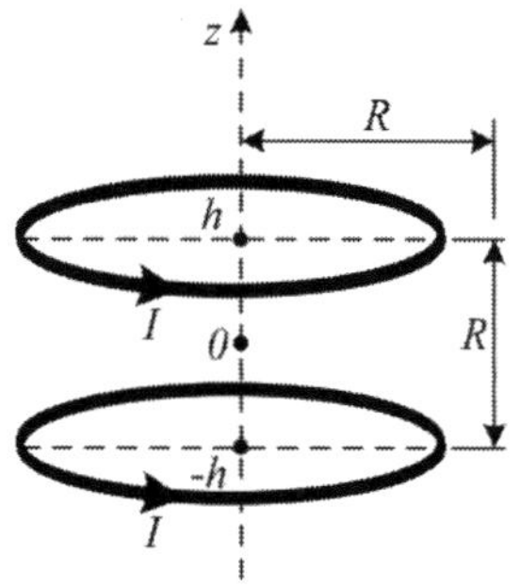

Figure 1. Helmholtz coils.

Axial field of two coils can be calculated by formula (2):

$$B(z) = \frac{\mu_0 NIR^2}{2}\left\{\left[R^2 + (z+h)^2\right]^{-\frac{3}{2}} + \left[R^2 + (z-h)^2\right]^{-\frac{3}{2}}\right\}. \tag{2}$$

For Helmholtz coils $h$is$R$/2, then formula (2) take the following form:

$$B(z) = \frac{\mu_0 NIR^2}{2}\left\{\left[R^2 + \left(z+\frac{R}{2}\right)^2\right]^{-\frac{3}{2}} + \left[R^2 + \left(z-\frac{R}{2}\right)^2\right]^{-\frac{3}{2}}\right\}. \tag{3}$$

Consider the successive derivatives of $B(z)$ in z for estimate the uniform of the field in the center of the coil system. A higher order of the first non-zero derivative at this point provides better uniform of the field in its vicinity.

Equation (3) can be expanded in a Taylor series in the variable z near zero:

$$B(z) = B(0) + \frac{1}{2}B^{(2)}(0)z^2 + \frac{1}{24}B^{(4)}(0)z^4 + \frac{1}{720}B^{(6)}(0)z^6 + \ldots. \tag{4}$$

Since the Helmholtz coils are arranged at a distance of radius from each other, the second derivative in the central of system becomes zero, i.e. $B^{(2)}(0) = 0$.

Thus, the change in the field ΔB when moving from the point $z = 0$ is given by (5) and has a fourth order of smallness:

$$\Delta B(z) = \frac{1}{24}B^{(4)}(0)z^4 + \frac{1}{720}B^{(6)}(0)z^6 + \ldots. \tag{5}$$

The magnetic field at the geometric center of the Helmholtz coils ($z = 0$) is determined from the expression:

$$B(0) = \frac{16}{5\sqrt{5}} \frac{1}{2} \frac{\mu_0 NI}{R}. \tag{6}$$

From the expressions (3) and (6) the formula for the relative nonuniformity of the magnetic field along the axis z is:

$$\delta_z = \frac{B(z) - B(0)}{B(0)} = \frac{5\sqrt{5}}{16} \left\{ \left[1 + \left(\frac{1}{2} + \frac{z}{R} \right)^2 \right]^{-\frac{3}{2}} + \left[1 + \left(\frac{1}{2} - \frac{z}{R} \right)^2 \right]^{-\frac{3}{2}} \right\} - 1. \tag{7}$$

Figure2 shows graph of the relative nonuniformity of magnetic field δ_z by the formula (7), where values of the z-axis coordinates are given the coil radius R.

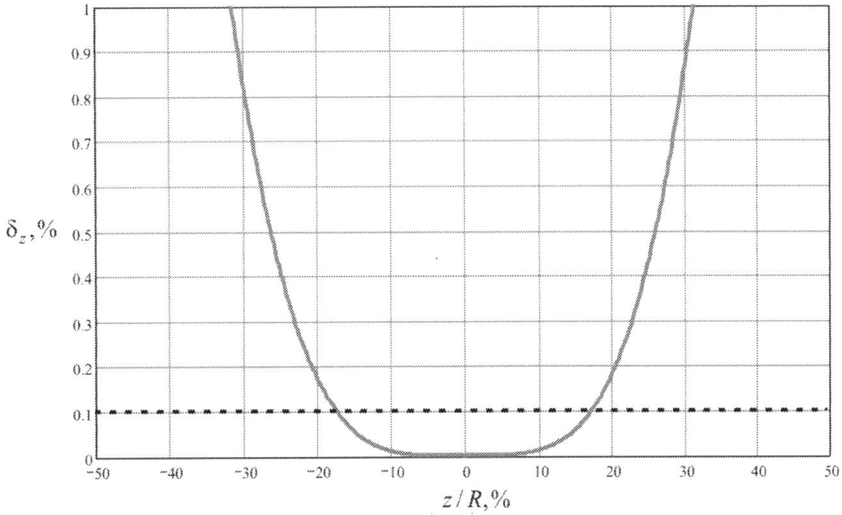

Figure 2. The nonuniformity$\delta_z(z/R)$ of the magnetic field generated by the Helmholtz coils.

Figure 2 shows that the magnetic field nonuniformity$\delta_z \leq 1$ % can be obtained at a distance from the geometric center $z/R<$ 31.4 %. If desired nonuniformity$\delta_z \leq 0.1$ % (shown by the broken line in Figure 2), the distance z/Rshould be less than 17.3 %.

By analogy with the calculation of magnetic field along the z axis to obtain expression of a cylindrical coordinate system for the axial component $B_z(z, r)$ of the magnetic induction at any point in the volume inside the Helmholtz coils:

$$B_z(z,r) = \frac{\mu_0 NI}{2\pi} \left[\begin{array}{l} \displaystyle\int_0^{\pi} \frac{R(R - r\cos\alpha)}{[R^2 + r^2 - 2Rr\cos\alpha + (z - R/2)^2]^{3/2}} d\alpha + \\ \displaystyle + \int_0^{\pi} \frac{R(R - r\cos\alpha)}{[R^2 + r^2 - 2Rr\cos\alpha + (z + R/2)^2]^{3/2}} d\alpha + \end{array} \right]. \tag{8}$$

Then axial component relative nonuniformity of magnetic induction at any point in the volume inside the Helmholtz coils is calculated by the formula:

$$\delta_z(z,r) = \frac{B_z(z,r) - B(0)}{B(0)}. \tag{9}$$

Calculation by formula (9) shows that the magnetic field nonuniformity$\delta_z(z, r) \leq 0{,}1$ % can be obtained at a distance from the geometric center z/R is less than 17.3 % and r/R is less than 15%.

In the paper [9] we described 8-coil system based on Helmholtz coils with $R = 55$ mm. Parameters of the 8-coil system given in Table 1.

Table 1. Parameters of the 8-coil system to create a uniform magnetic field.

Number coil	The number of turns	The distance from the coil center to the origin, (mm)
1; 5	10	13,5; –13,5
2; 6	12	42,5; –42,5
3; 7	17	80,5; –80,5
4; 8	34	144,5; –144,5

3 Modeling axial 8-coil system

Software COMSOL Multiphysics 4.4 for modeling 8-coil systemis used. Figure 3 shows the geometry of the coils system created in COMSOL.

The calculation of the magnetic field produced in the module *mf* as only it allows using the section *Multi-Turn Coil Domain* coils for modeling coils in 3D. The coils have closed geometry with a circular cross-section perpendicular to the axis *z*. *Circular type* of coil was chosen. The direction of current flow in this case is modeled by specifying the border (reference edge), along which the current flows.

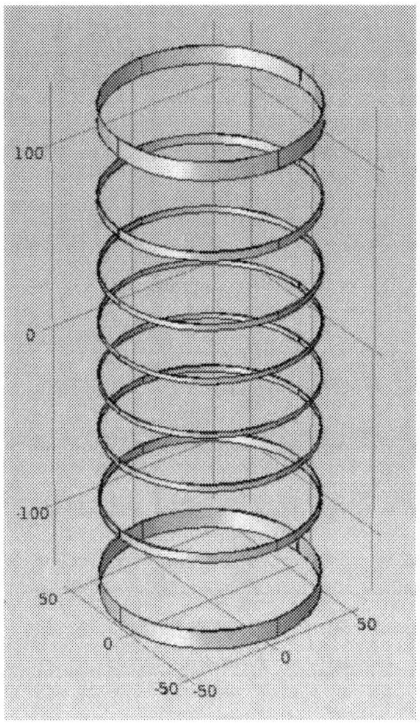

Figure 3. The axial 8-coil system in COMSOL.

Modeling results of the coils system at the DC current of 0.001 A are shown in figure 4 and figure 5.

$$\delta_{z.cae}(z) = \frac{B(z) - B(0)}{B(0)} \cdot 100\,\% \qquad (10)$$

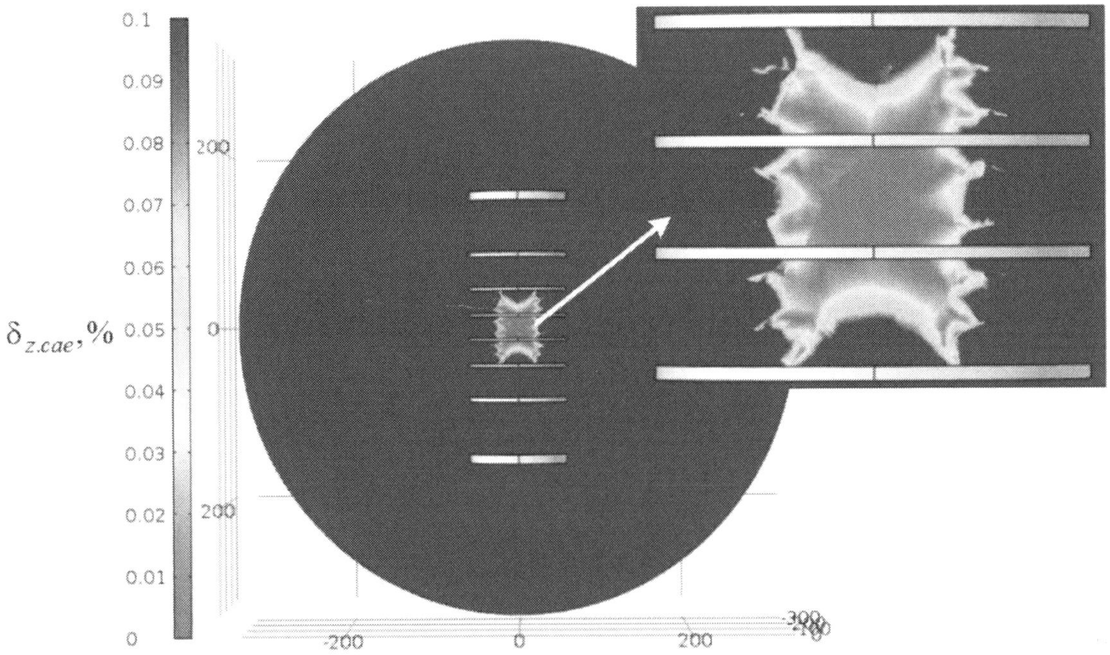

Figure 4. The distribution of the magnetic field in the axial direction.

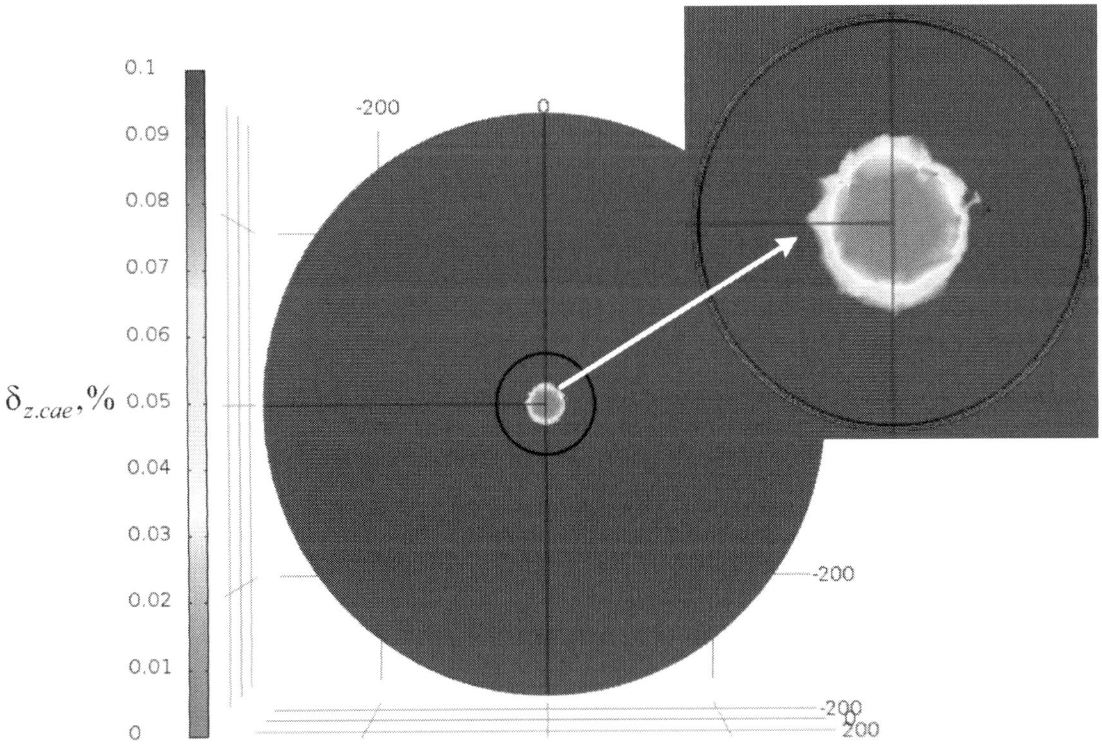

Figure 5. The distribution of the magnetic field in the radial direction.

Nonuniformity of the magnetic field was calculated by formula (10) and is shown in Figure 6.

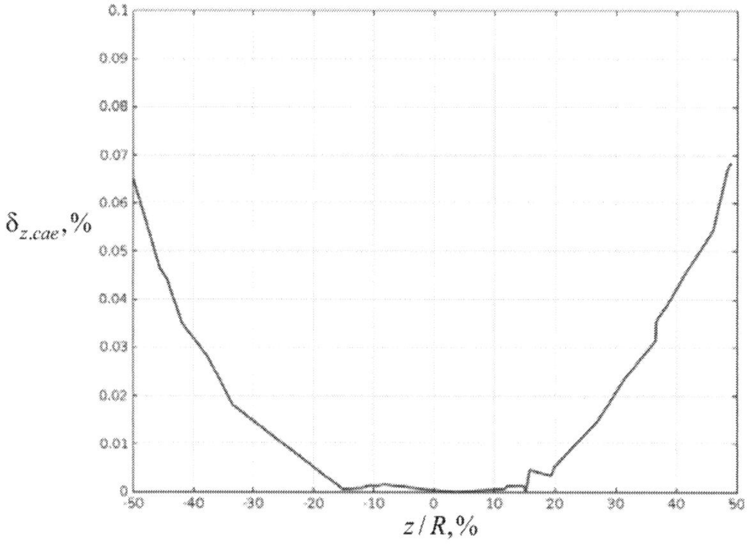

Figure 6. The nonuniformity of the magnetic field.

4 Conclusion

A standard of the DC magnetic field requires for test, calibration and graduation of sensors of magnetic fields and magnetometers based on them.

As a result of work modeling 8-coil system to generating uniform magnetic field with increased volume in comparison Helmholtz coils was described. Parameters of axial 8-coil system for generating uniform magnetic field are described. Modeling results of axial 8-coil system showed that the uniformity of the field production by a system of 8 coils of 55 mm radius is not less than 0.07 % at distances up to $R/2$ from the center of the system.

The obtained results show that such a system can be used to test, calibrate and graduate magnetic sensors and magnetometers based on them.

Acknowledgement

The work has been supported in part by the grant of the RFBR, research project No. 15-08-01007.

References

1. R.A. Schill, and H. Karin, *Review of Scientific Instruments* **72**, 2769 (2001)
2. Po Gyu Park, et al., *CPEM*, 456 (2008)
3. Po Gyu Park, et al., *CPEM*, 312 (2010)
4. D. Lorrain, D. Corson (Freeman, San Francisco, 1979)
5. J. D. Kraus (McGraw–Hill, New York, 1973)
6. P.M. Fishbane, S. Gasiorowicz, S.T. Thornton (Prentice Hall, New York, 1996)
7. M. Hosoya, E. Goto, *Review of Scientific Instruments* **62**, 2472 (1991)
8. V.E. Baranova, P.F. Baranov, *Dynamics of Systems, Mechanisms and Machines* (2014)
9. V.E. Baranova et al., Measurement Techniques **58**, 550 (2015)

MATEC Web of Conferences 48, 03002 (2016)
DOI: 10.1051/matecconf/20164803002

Non-contact calibration technique of ultrasound tomography system for complex shaped objects

German Filippov[1,a], Vadim Zhvyrblya[1], Svetlana Sharavina[1] and Yana Salchak[1]

[1]National Research Tomsk Polytechnic University, 634050 Tomsk, Russia

Abstract. Robot-assisted ultrasound tomography is an up-to-date technology that provides effective quality assurance of complex shaped objects. However, this method of control requires appropriate calibration of the equipment. At present, calibration is a difficult time-consuming procedure. This paper discusses novel approach – using non-contact calibration technique. The proposed method is based on the preliminary received 3-D reconstruction results. It is proved that non-contact calibration of ultrasound tomography system is more applicable and provides reliable measurement results.

1 Introduction

Nowadays ultrasound techniques are one of the most valuable methods of quality assurance. The main advantage of ultrasonic testing compared with other methods is ease of its implementation. It is also important that this type of control is applicable for the variety of materials, such metals and nonmetallic structures. Moreover, ultrasound method of control has high rate, low cost and is absolutely safe as compared to X-ray testing).

Ultrasonic testing techniques have improved significantly for the last several years. Usually it used to be manual instruments that were not able to create three-dimensional images of defects. At present, there are novel systems of ultrasound tomography that enable highly accurate reconstruction of the defects of a controlled object.

Recent development of additive technologies resulted into new requirements for ultrasonic systems. This is due to the fact that objects of control became more complex. For example, objects that are produced by 3-D printing technologies are usually not linear and have specific design. Hence, it is necessary to develop a quality assurance system for complex shaped objects.

2 Experimental Procedure

Robot-assisted ultrasound tomography system consists of six-axes scanner, immersion tank and electronic unit (see Fig.1). The major challenge for this type of testing systems is reliable estimation of relative position of the scanner and the controlled object. Today it can be determined by contact methods.

[a] Corresponding author : ilippovga@gmail.com

An operator of the system is supposed to contact sides of the object with the special calibration tool. This procedure is time-consuming, has low accuracy and is not always possible due to the necessity of direct contact with an object.

Current algorithm of generation of the scanning path along the object of control consists of several steps. First, special instrument with a sharp nozzle is assembled at the scanner. Then, this instrument is placed at the particular points of the object for coordinates estimation (given to the length of the instrument). The determined coordinates are saved by the means of the software and compared with the chosen points at the object surface (typically they are angles). The whole process is to be carried out very accurately as the smallest mistake can lead to re-starting of calibration measurements.

Therefore, appliance of 3D- scanner system was proposed as a solution for the reason that it determines the coordinates of all points of an object.

The system consists of three main parts: instrumental unit; immersion tank and scanner.

The instrumental unit includes: PC, power supply, scanner controller, ultrasound electronics unit.

Industrial robot is applied as an automatic scanner. It was developed by German company KUKA, model number KR10R1100 six WP. The scanner performs positioning of the transducer on the given distance from the controlled object.

Figure 1. Ultrasound tomography system.

Ultrasound electronics unit performs functions connected with transmitting and receiving acoustic signals, multiplexing of channels, digitizing of the signals, data transfer and processing (see Fig.2). This unit has the following technical parameters:

- 128 multiplexed channels (16 parallel channels)
- capacity of 200 MB/s
- voltage of transmitting module: from -50 to -180 V
- the minimum of input sensitivity: 100 mkV
- the maximum of input voltage: 10 V
- the resolution of AD converter: 14 bit
- sampling frequency: 20 MHz
- AD buffer: 64 000 measurements.

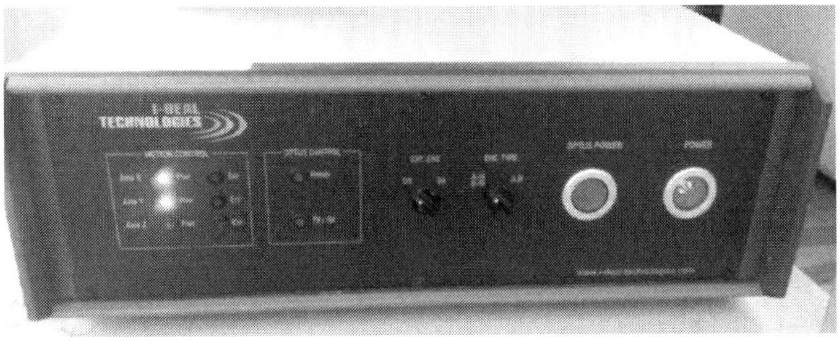

Figure 2. Ultrasound electronic unit.

It is sufficient to have only three points of both the known object and the object of control that fall inside the scanning area for the estimation of the controlled object in accordance with the scanner coordinate system.

A special calibration disk was designed for this purpose and then manufactured by 3-D printing (see Fig.3).

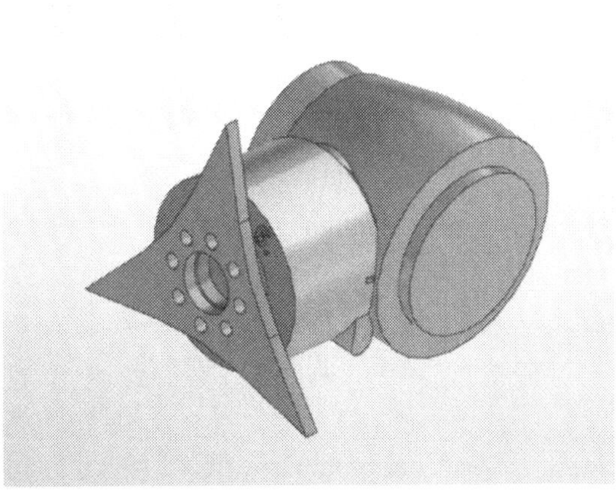

Figure 3. Calibration disk.

The developed calibration disk is rigidly fixed on the scanner flange. The scanner is guided to the controlled object in such a way that three angles of the disk and three points of the controlled object are located inside the scanning area. The point cloud is formed as a result of the scanning process (see Fig.4).

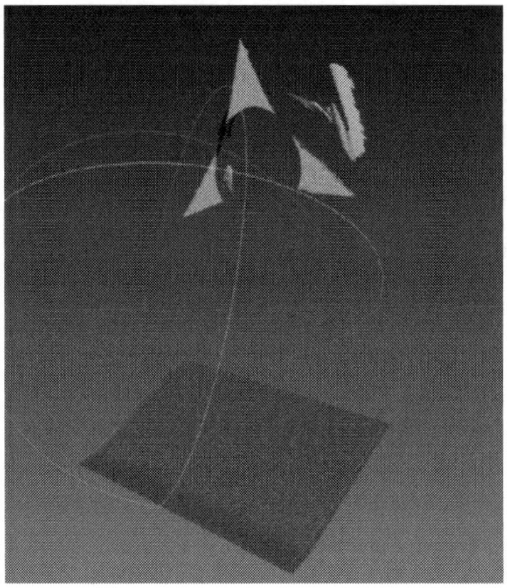

Figure 4. The point cloud.

It is possible to determine the location of an object in accordance with the scanner coordinate system by combination of the model of the object and the scanner with the point cloud. As a result, the matrix is formed. It provides position of the object regarding to the scanner location. Calculation can be made in accordance with [1] or can be realized by PCL libraries.

The transformation matrix for the given measurement was formed in accordance with [1]:

$$\begin{bmatrix} -0.853 & 0.070 & -0.515 & 1188.343 \\ -0.069 & -0.997 & -0.022 & 155.978 \\ -0.516 & 0.016 & 0.856 & 710.207 \\ 0 & 0 & 0 & 1 \end{bmatrix}$$

The matrix formed as a result of contact calibration method is given below:

$$\begin{bmatrix} -0.875 & 0.064 & -0.611 & 1188.525 \\ -0.061 & -0.985 & -0.032 & 155.478 \\ -0.523 & 0.028 & 0.835 & 710.951 \\ 0 & 0 & 0 & 1 \end{bmatrix}$$

Comparison of the matrixes reveal that both calibration methods provide the same result.

An object made of composite materials with artificial defects was chosen as an experimental sample (see Fig. 5). Ultrasound testing was chosen as the method of control [2].

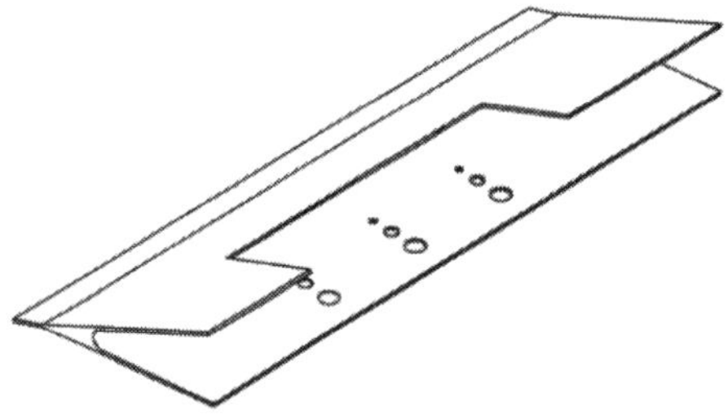

Figure 5. Carbon fiber experimental sample with artificial defects.

The ultrasound tomography system [3] was calibrated by contact and non-contact methods during the experiment (Figures 6 and 7). The most accurate results of control were supposed to determine the most efficient calibration method.

3 Conclusion

Comparison of the obtained results demonstrated that both calibration methods provide quality control assurance with the same level of accuracy. That is why the proposed method can be applied as a valid calibration technique for ultrasonic tomography systems. It can be used for nondestructive testing of the objects in critical industrial areas such as aerospace and aviation. Moreover, it is evident that the non-contact method of calibration is less time-consuming technique that enables effective nondestructive quality control.

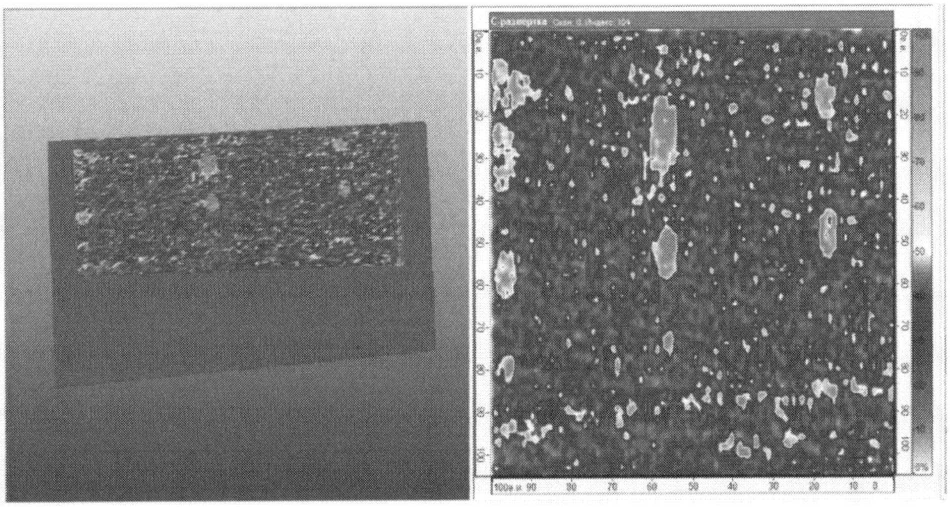

Figure 6. Results of control by contact calibrated system.

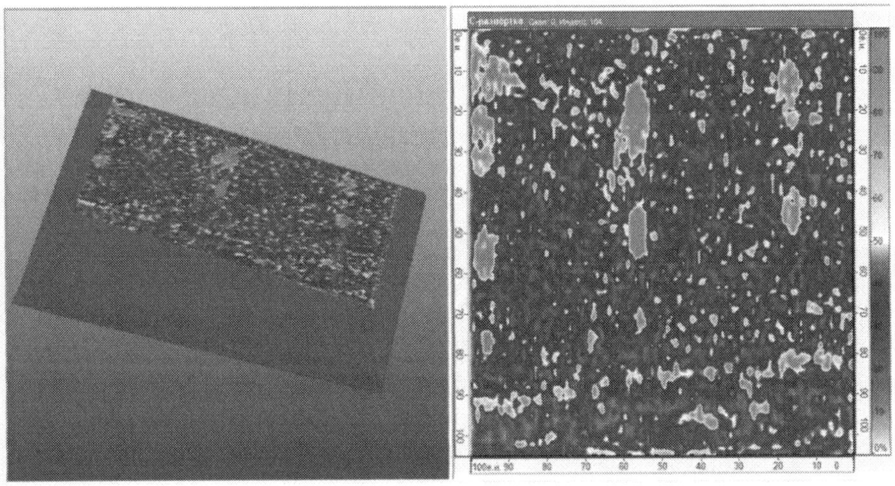

Figure 7. Results of control by non-contact calibrated system.

Acknowledgement

Research has been conducted with the financial support of Federal Target Programme, project unique identifier is RFMEFI57514X0048.

References

1. V. Borikov, O. Galtseva, G. Filippov, IOP Conf. Ser.: Mat. Sc. and Eng. **671**, 012014 (2016)
2. Y. Salchak, A. Bulavinov, R. Pinchuk, A. Lider, I. Bolotina, D. Sednev, PP **70**, 484 (2015)
3. K. Kvasnikov, A. Soldatov, I. Bolotina, M. Krening, A. Potapenko, Rus. J. of NDT **49**, 625 (2013)

MATEC Web of Conferences 48, 03003 (2016)
DOI: 10.1051/matecconf/20164803003

Universal digital strain gauge measurement system of aeroelastic deformation development

Anton Pavlov[1, a] and Alexander Arefiev[1]

[1]Saint-Petersburg University of Aerospace instrumentation, 190000 Saint -Petersburg, Russia

Abstract. This article presents description of the universal digital strain gauge system developed to measure the static and dynamic aeroelastic deformations of elasticity-scale models during the tests in aerodynamic tube and during in-flight tests of an experimental air vehicles. The main requirements for such devices are small size and possibility of operation in a wide temperature range. The article considers the dependence of zero offset from temperature. Functional diagram block and logic diagram of the build system are shown.

1 Introduction

Today the great attention is paid to concept investigation of an active aeroelastic wing. This concept allows to improve maneuvering characteristics for a wide range of speed, angles of attack, heights and at the same time reducing weight of the aircraft [1].

The synthesis of the control system within the concept of aeroelastic wing necessarily requires an application of the principle of integrated design, in which all the links of the chain: design – aerodynamics – strength are related. Control system of aeroelasticity requires high attention to the accuracy of determination of parameters of the aircraft in the wind tunnel tests, including elasticity-scale models. Characteristics of elasticity of the studied objects during wind tunnel and in-flight tests are basically defined with the help of strain gauge measurements.

This article presents the concept of universal digital strain gauge system that can be installed on elasticity-scale models for wind tunnel tests and in experimental air vehicles [2].

Thus, the main system requirements are: miniaturization of its components, and possibility of its operation in a wide temperature range. In addition, measurement components of this system could be connected by wired connection or/and wireless.

2 Description of system

The described strain gauge system allows taking date simultaneously from not less than 50 strain gauge measuring units, and they could be connected by wired and/or wireless way.

[a] Corresponding author : sa807@bk.ru

2.1 Structure of system

System consists of: conversion unit USB-RS485; radio transmit-receive unit; signals distribution unit; strain measuring force unit; strain measuring deformation unit; extender interface.

All system components have a standard interface RS485 and power 6 to 25V, and input and output interface that allows to connect blocks in parallel to signals distribution unit or sequentially to each other, and also through the radio transceiver unit. Figure 1 shows an example of connecting the measuring units in a common measurement system and collection of information that can be installed on the investigated object.

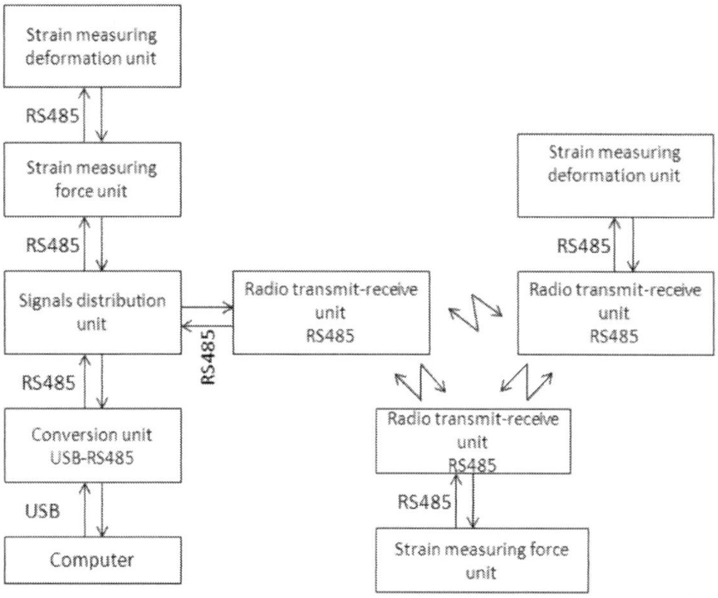

Figure. 1. Structural diagram of connection blocks in a single system designed for installation on elasticity-scale model.

2.2 Strain measuring force unit

Strain gauge in this unit is used to measure the force. As a sensitive element it uses strain gauge measuring pressure with strain-gage bridge. Structural diagram of the strain measuring force unit is shown in Figure 2. This unit is made with modular design and allows to use a large variety of strain gauges [3].

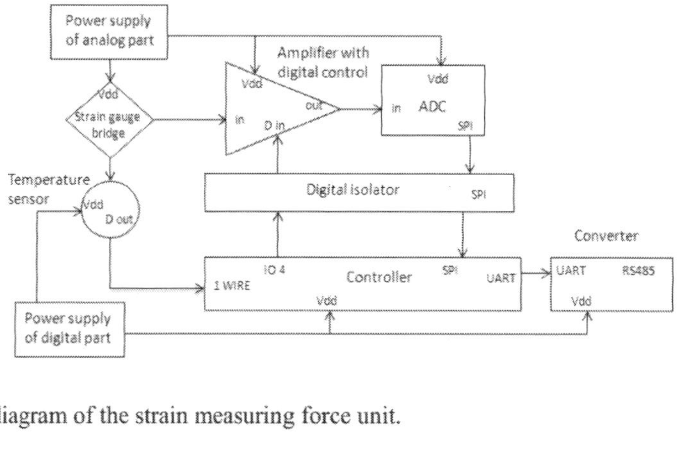

Figure 2. Structural electric diagram of the strain measuring force unit.

Strain measuring force unit can be divided into two parts: analog and digital. Analog part consists of: power supply of analog part; strain gauge bridge; amplifier with digital control; ADC. There is a low-pass filter at the output of the strain gauge. The amplifier converts the analog signal from the strain gauge to the signal from 0 to 3.3 V. Amplifier is equipped with a digital control zero offset and scale factor. ADC converts analog signal to digital and transmits it to the controller by SPI interface.

Digital part consists of: controller; power supply of digital part; temperature sensor; digital isolator; converter USB-RS485.

Digital isolator is used for galvanic isolation of analog part from digital, to reduce the noise effect generated by digital parts to the power supply of analog part. The temperature sensor measures the temperature of the strain gauge and circuit pack. It should be noted that each strain measuring force unit is previously calibrated in a climatic chamber at the required temperature range. The aim of calibration is the determination of scaling factors of the strain gauge and zero offset in the operating temperature range.

Thus, the temperature sensor serves to eliminate the influence of temperature on measured parameters of the strain gauge. The controller reads the data, filters, processes the request, and sends to the consumer via converter USB-RS485.

The firmware of the strain measuring force unit provides: setting of the amplifier; reading of ADC parameters; reading the data from the temperature sensor; filtering the data; conducting of compensation of bias and scale temperature factor; receiving and processing external commands; recording calibration data. Such implementation simplifies the work with strain gauge sensor and reduces the amount of post-processing.

2.3 Strain measuring deformation unit

Strain measuring deformation unit is used to measure elastic deformations, arising under tension-compression, bending and torsion. As a sensitive element in this unit is used strain-gauge plate mounted by special glues. If the test objects are held under high humidity for a long time, the sensitive elements are protected by a moisture-resistant coatings. The electronic part of this is completely similar to the strain measuring force unit. The main difference in comparison with strain measuring force unit is using another type of sensitive element, method of its calibration and processing sensor data. This unit also enables the connection a wide variety of analog sensors.

2.4 Measuring oscillation unit

One of the most important areas of investigations of dynamic characteristics of aeroelastic oscillations is their bodies and control systems. These studies enable to obtain extremely valuable information about the effectiveness of combating with flutter and buffeting. The basis of the proposed strain-measuring units can be used as the unit for measuring the frequency characteristics of the test objects. For this purpose, it is necessary to increase the frequency of removal of parameters of strain measuring force unit, connect the accelerometer as a sensitive element and change its firmware.

3 The layout of the strain measuring force unit

The layout of the strain measuring force unit is made of elements represented in the Figure 2. The layout is connected to personal computer through USB, for transmission of processed force and temperature data for further processing in it. The layout of the strain gauge measuring force unit is shown in Figure 3. Making layout pursuit the aim to test the circuitry of this block, as well as studies of the temperature effect on zero offset of the strain gauge, and to determine the proper allocation of the temperature sensor. Figure 4 shows a graph of zero offset from temperature. It should be noted, that the layout is designed in such a way that allows connecting a wide range of strain gauges.

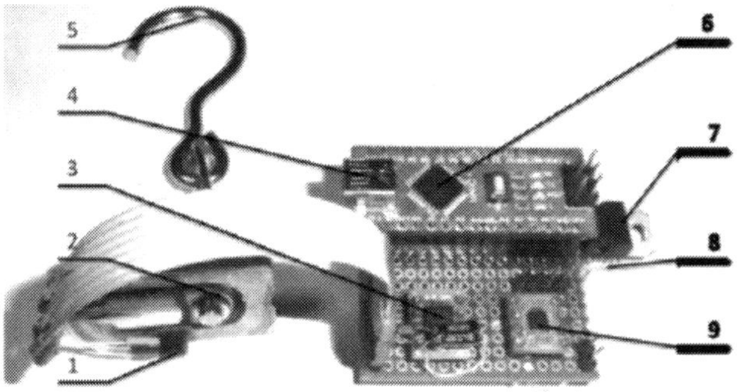

Figure 3. The layout of the strain gauge measuring unit of force. 1 — Temperature sensor, 2 — strain gauge, 3 — amplifier, 4 — USB connector, 5 — hook of strain gauge, 6 — controller, 7 — analog power supply, 8 — digital isolator, 9 — ADC.

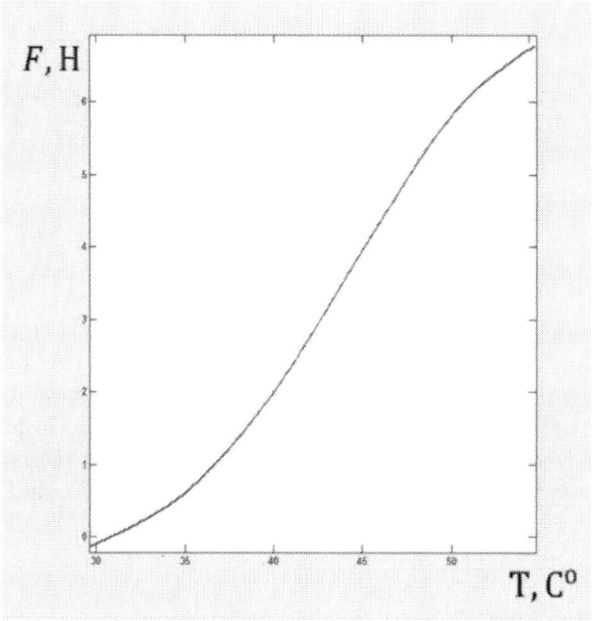

Figure 4. A graph of the dependence of zero offset from temperature.

4 Conclusion

This article presents the concept of a universal digital strain measurement system of aeroelastic deformations. Structural system diagram and layout of the strain gauge measuring force unit were introduced. In addition, a temperature effect on zero offset of the strain gauge was shown.

The prototyping of components of the system is going to reduce the dimensions of each unit. Thus, it will be possible to ensure the installation of the proposed system on elasticity-scale models, which are used in wind tunnel tests and experimental air vehicles.

References

1. Campanile, L. F. and Sachau, D. J. Intell. Mater. Syst. Struct. **10**, 215 (2000)
2. J. R. Wright, J. E. Cooper, *Introdaction to Aircraft Aeroelasticity and Loads*, ISBN 978-0470-85840-0 (2007)
3. M. Jafaripanach, B. M. Al.-Hashimi, N. M. White, IEEE International Symposium on Circuits and systems **4**, IV 752 (2003)

MATEC Web of Conferences 48, 03004 (2016)
DOI: 10.1051/matecconf/20164803004

Application of frequency-domain algorithms in ultrasound imaging of composite materials

Dmitry Dolmatov[1,a] and Vladislava Abramets[1]

[1] *National Research Tomsk Polytechnic University, 634050 Tomsk, Russia*

Abstract. Non-destructive ultrasonic testing of composite materials requires imaging algorithms which could be applied for anisotropic structures. At present, time-domain techniques are used for this purpose. However, it is evident that this method has lower computational speed compared with frequency-domain algorithms. This paper discusses advanced approach off frequency-domain techniques for multilayer composite materials inspection. The proposed method is based on multi-layer omega-k (MULOK) ultrasound frequency-domain imaging algorithms. It is shown that MULOK algorithms enable highly accurate reconstruction within a higher computational speed.

1 Introduction

Composite materials are used in aerospace engineering extensively. Application of these structures requires nondestructive testing that enables to reduce the risk of accidents. One of the methods, which could be used for this purpose, is ultrasonic nondestructive testing [1]. Furthermore, advanced ultrasonic imaging algorithms allow accurate determination of defect and its size, shape and location in the controlled object. That information is to be used for reliability assessment of composite structures. However, ultrasonic visualization and image reconstruction of anisotropic structure of composite materials is one of the major challenges due to the acoustic properties changes within the direction of the wave.

2 Ultrasound imaging algorithms

Synthetic aperture focusing technique (SAFT) used to be the first ultrasound imaging algorithm [2]. This algorithm proposes image reconstruction according to the following formula:

$$I(x,z)=\left|\sum_{tr=1}^{N} P(\tau, x_{tr})\right| \tag{1}$$

$P(\tau, x_{tr})$ is a value of ultrasound field in the position of ultrasound transducer x_{tr} at the time τ. Correlation between τ and coordinates of the image x and z is following:

[a] Corresponding author : dolmatovdo@tpu.ru

$$\tau = \frac{2\sqrt{(x-x_{tr})^2+z^2}}{C_l} \tag{2}$$

C_l is the speed of longitudinal ultrasonic waves in the specimen. This algorithm can be applied only in the case of isotropic materials. Further, SAFT was adjusted for multilayer structures imaging [3] that could be used in ultrasonic imaging of composite materials. However, it is known that frequency-domain algorithms have better computational speed than time-domain algorithms [4]. Frequency-domain algorithms can be implemented in two ways. First algorithm was proposed by Gazdag [5] and was used in seismology. This technique is known as phase shift migration (PSM) algorithm. Afterwards, it was improved and successfully applied for ultrasonic nondestructive testing [6]. The imaging algorithm implies slice by slice reconstruction of the image according to the following correlation:

$$I(x,z+\Delta z) = \int_{-\infty}^{+\infty} (\sum_{\omega} P(\omega,k_x,z+\Delta z))\, e^{ik_x x} \tag{3}$$

Each slice of the image is a result of inverse Fourier transformation of sum $P(\omega,\mathrm{k}_x,z+\Delta z)$ along the wave number k_x. Function $P(\omega,\mathrm{k}_x,z+\Delta z)$ is the extrapolated wave field, which is derived into two steps. First, 2-D Fourier transformation of registered wave field is conducted:

$$P(\omega,k_x,z) = \frac{1}{4\pi^2}\int_{-\infty}^{+\infty}\int_{-\infty}^{+\infty} P(t,x,z)\, e^{-i(k_x x-\omega t)}dx\,dt \tag{4}$$

Then, the wave field is extrapolated to the depth $z+\Delta z$:

$$P(\omega,k_x,z+\Delta z) = P(\omega,k_x,z)\cdot e^{(i\cdot k_z(\omega,k_x)\cdot\Delta z)} \tag{5}$$

Correlation between wave numbers k_z, k_x and temporal frequency ω is following:

$$k_z(\omega,k_x) = \sqrt{\frac{4\omega^2}{C_l^2}-k_x^2} \tag{6}$$

Stolt [7] proposed second approach.This method suggests image reconstruction according to the following correlation:

$$I(x,z) = \iint (\frac{C_l}{2}\frac{k_z}{\sqrt{k_x^2+k_z^2}})P(k_x,k_z)e^{-ik_x x-ik_z z}dk_x dk_z \tag{7}$$

The main part of this algorithm is the change of the variable. Temporal frequency w is changed to the wave number k_z:

$$w(k_x,k_z) = \frac{C_l}{2}\cdot\sqrt{k_x^2+k_z} \tag{8}$$

This algorithm can be applied only for isotropic materials. However, it can be adapted for multilayer structures testing by combination this algorithm with PSM technique [8]. Initially, first layer is reconstructed by using (7). After that, wave field is extrapolated according to (5) to the depth where layer with new properties starts. The last step is wave field extrapolation that is used in Stolt algorithm for reconstruction of second layer and so forth. Application of this approach in ultrasound imaging is known as Multilayer omega-k (MULOK) algorithm [4].

3 Experimental procedure

A number of simulated processing was performed in Matlab R2009b (The MathWorks, Natick, MA) in order to compare two algorithms. The simulations were made on a dual-core 2MHz, 8 Gb RAM computer. Simulated material consists of two layers. First layer was made of copper, second layer was made of aluminum. The thickness of both layers was 25 mm. The length of simulated specimen was 40 mm. The simulated specimen is shown in the Figure 1.

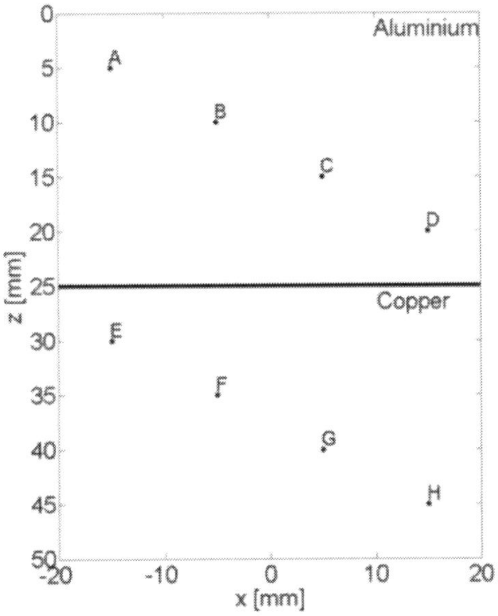

Figure 1. The simulated specimen.

The transducer which was used in this simulation had 5 MHz frequency and diameter of 3 mm. Sampling frequency was 50 MHz. The step of the transducer was changed for each simulation. Four simulations were made for each algorithm. The step values were 1, 0.5, 0.2 and 0.1 millimeters. The results of simulations are shown in the Figure 2 for PSM and in the Figure 3 for MULOK.

The results of the simulation were analyzed in order to estimate the computational speed and the quality of the reconstructed image. First parameter can be defined by measuring elapsed time for each simulation. Correlation between calculation time and number of steps in each simulation is presented in the Figure 4. Quality of the reconstructed image can be evaluated according to the following relation [9]:

$$P = \frac{A^{-6dB}}{\lambda_c^2} \qquad (9)$$

A is an area where a point spread function is greater than -6dB down from its local maximum value whereas λ_c is a transducer center wavelength. This value was determined for both algorithms at the step 0.1 mm. The results are presented in the Table 1. The difference between values for both algorithms was derived by using the following correlation:

$$D = \frac{P^{PSM} - P^{MULOK}}{P^{PSM}} \cdot 100\% \qquad (10)$$

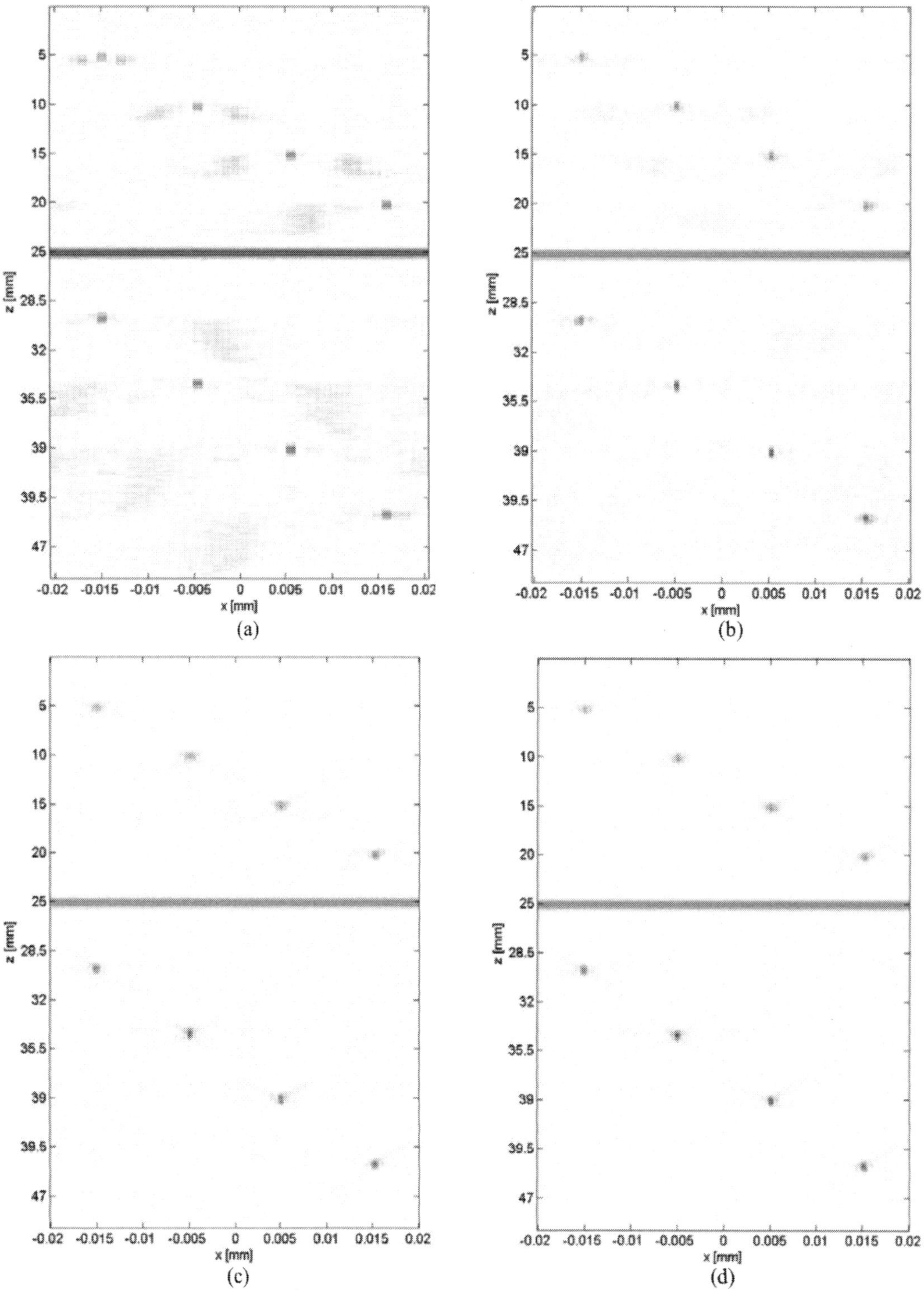

Figure 2. The results of simulations by using PSM algorithm: (a) 1 mm transducer step, (b) 0.5 mm transducer step, (c) 0.2 mm transducer step, (d) 0.1 mm transducer step.

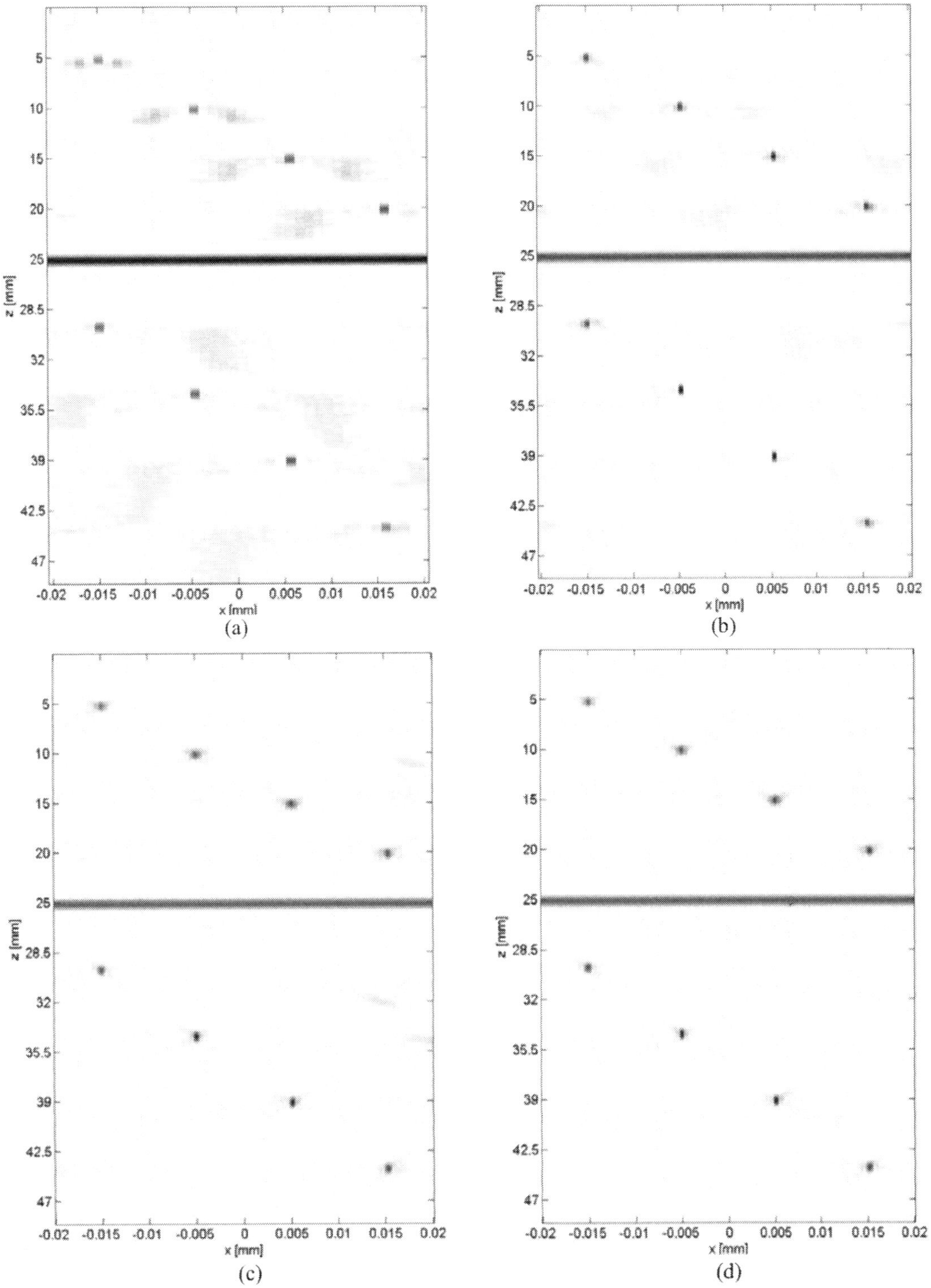

Figure 3. The results of simulations by using MULOK algorithm: (a) 1 mm transducer step, (b) 0.5 mm transducer step, (c) 0.2 mm transducer step, (d) 0.1 mm transducer step.

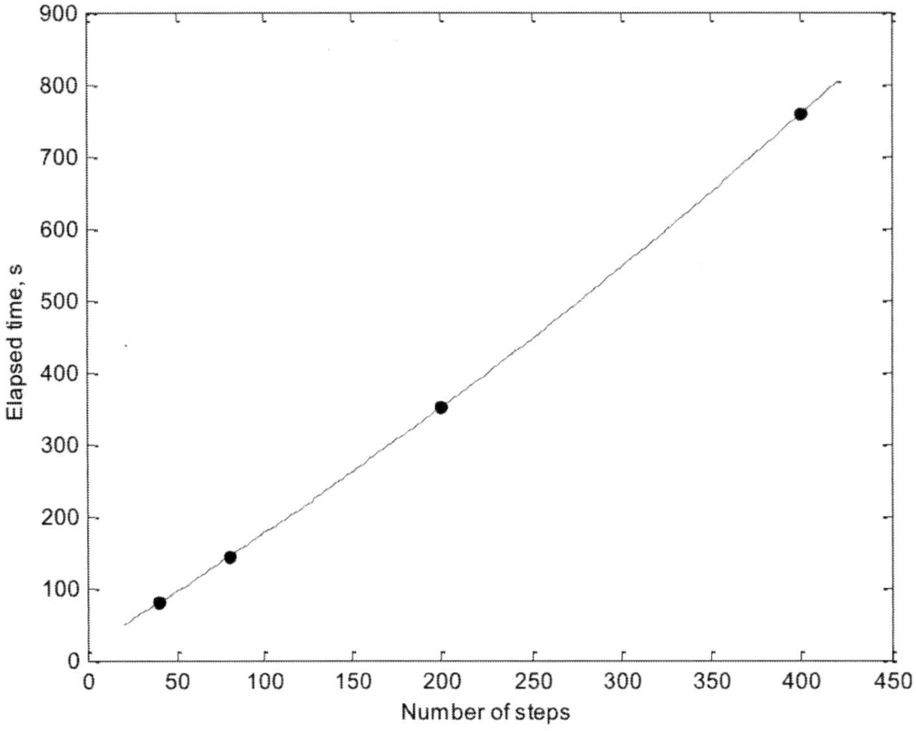

(a)

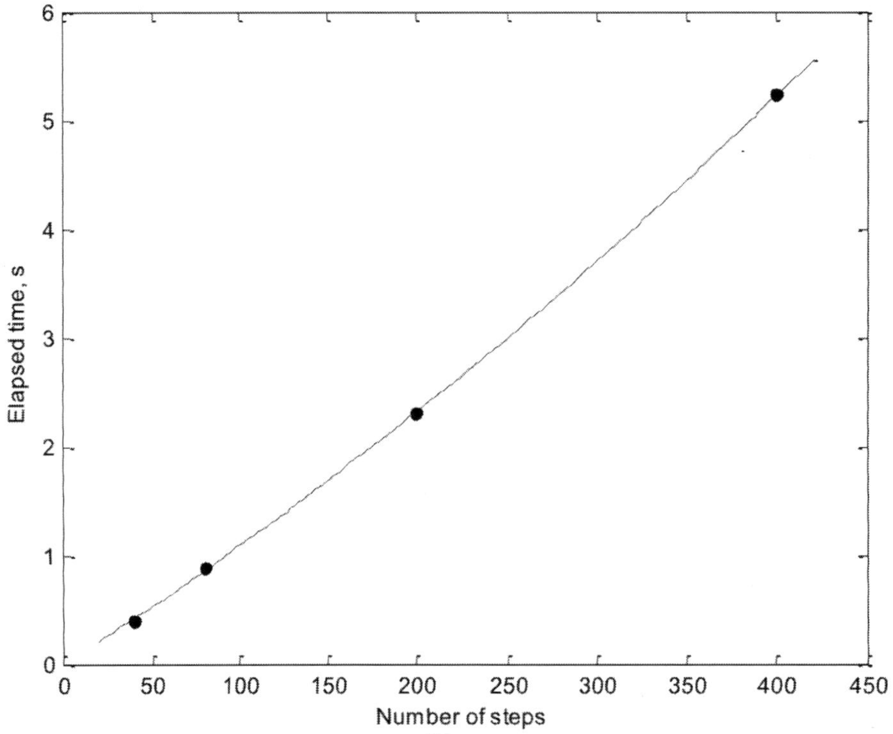

(b)

Figure 4. The correlation between calculation time and number of steps in each simulation: (a) PSM, (b) MULOK.

Table 1. Performance indicators of MULOK and PSM algorithms.

Point	MULOK	PSM	Difference, %
A	2,13	2,18	2.1
B	2,17	2,18	0.5
C	2,22	2,22	0
D	1,78	1,91	7.6
E	1,91	1,91	2.9
F	1,40	1,49	6.1
G	1,46	1,51	3.9
H	1,89	1,90	0.7

Comparison of the obtained results demonstrates that MULOK algorithm is faster than PSM. Computational time of imaging algorithm is very important in ultrasound imaging due to the fact that in nondestructive testing it is necessary to work with huge set of data. Also, Table 1 represents that the quality of the images for both methods is almost the same. That is why, experimental results reveal that MULOK algorithm enables effective ultrasound imaging for multilayer materials within a higher computation speed.

4 Conclusion

This study represents that frequency-domain ultrasound imaging algorithm can be used for anisotropic materials, which consist of isotropic layers. There were two-dimensional case and two layers case studied separately. However, the same simulation can be made for three-dimensional case and multilayer material if it is required. The obtained results demonstrated that MULOK algorithm can be successfully applied in practice. However, this algorithm is not to be used for ultrasound imaging of all types of composite materials due to the diversity of its structure and properties. It is possible to apply MULOK algorithm for composite materials which are meet the conditions of simulations.

Acknowledgement

Research has been conducted with financial support of the Federal Target Program, project unique identifier is RFMEFI57514X0048.

References

1. G. Scott, C. M. Scala, NDTi **15.2**, 75 (1982)
2. S. R. Doctor, T. E. Hall, L. D. Reid, NDTi **19.3**, 163 (1986)
3. M. H. Skjelvareid, Y. Birkelund, *ASME 2010 Pressure Vessels and Piping Division/K-PVP Conference* (2010)
4. M. H. Skjelvareid, UFFC **58.5**, 1037 (2011)
5. J. Gazdag, G. **43.7**, 1342 (1978)
6. T. Olofsson, UFFC **57.11**, 2522 (2010)
7. R. H. Stolt, G. **43.1**, 23 (1978)
8. Y. C. Kim, R. Gonzalez, J. R. Berryhill, G. **54.3**, 319 (1989)
9. A. J. Hunter, B. W. Drinkwater, P. D. Wilcox, UFC **55.11**, 2450 (2008)
10. K. G. Kvasnikov , A. I. Soldatov , I. O. Bolotina , M. K. Krening, A. A. Potapenko, RJNT **49.11**, 625 (2013)
11. D. Sednev, A. Lider, D. Demyanuk, M. Kroening, Y. Salchak, PP **70**, 505 (2015)

MATEC Web of Conferences **48**, 03005 (2016)
DOI: 10.1051/matecconf/20164803005

Ultrasonic tomography of complex shaped carbon fiber composites

Vadim Zhvyrblya[1,a], German Filippov[1], Elizaveta Sedanova[1] and Yana Salchak[1]

[1]*National Research Tomsk Polytechnic University, 634050 Tomsk, Russia*

Abstract. Non-destructive quality assurance of complex shaped objects is one of the major challenges in the variety of industrial areas. Robot-assisted ultrasound tomography is one of the common methods used for this purpose. This study proposes a tomography based on the preliminary received 3-D reconstruction results. The obtained experimental details represent the applicability of the suggested method to ensure effective quality control.

1 Introduction

The rapid development of aerospace industry caused commensurate growth in composite industry. The geometry of composite structures has become more complex, that, in its turn, caused limitations of automatic nondestructive testing procedures. The most convenient in this case are robotic systems of control that enable accurate positioning of the transducer in accordance with the shape of an object and its surface normal.

However, implementation of a robotic system of control requires relevant 3-D model of the controlled object and precise location of the object according to the coordinate system of the equipment. One of the ways to create a 3-D model is three-dimensional reconstruction based on structured light techniques.

2 Digital Focus Array

At present, ultrasound testing methods are usually based on the registration of the echo-signal scattered from the back wall and inner flaws. The most common method is that implies the application of a single-channel transducer. That kind of transducer consists of one transmitting element and one receiving element. These two elements can be combined into one as well. In that case single-channel transducer emits acoustic wave and then receives the scattered signal.

However, the approach of single-channel transducers has several disadvantages compared with multi-channel transducers. For example, it is more time consuming and does not provide accurate results of the measurement. One of the testing methods where multi-channel transducers (phased array transducer) are applied is Digital Focus Array (DFA) technique.

The difference between single-channel testing method and DFA is the procedure of obtaining the data. Digital Focus Array Technique is applied with only one phased array element sending out a

[a] Corresponding author : vzhrb3001@gmail.com

signal, the rest elements are receivers. The signal can be considered as a wavelet that is generated by one individual element of the array. The phased array is moving on the fixed distance from controlled object during the inspection. Received time signals are saved and then form the matrix of units. This matrix is a basis for the further image reconstruction. The time of flight is evaluated for each unit. The necessary characteristics for image reconstruction are calculated according to the following correlation:

For the element with *x,y,z* coordinates the image reconstruction parameter is:

$$A(x, y, z)= \left| \sum_{x_b=x_{b1}}^{x_{b2}} \sum_{x_c=x_{c1}}^{x_{c2}} P_e(t_{bc}) \right| \quad (1)$$

$A(x,y,z)$ corresponds to the transparency of the image in x,y,z coordinate system. The high frequency echo-signals are summed up. Index "e" corresponds to one of elements that generate signals: transmitter and receiver. The time is correlated with the point of the image (x, y, z) in accordance with the location of ultrasonic transducers:

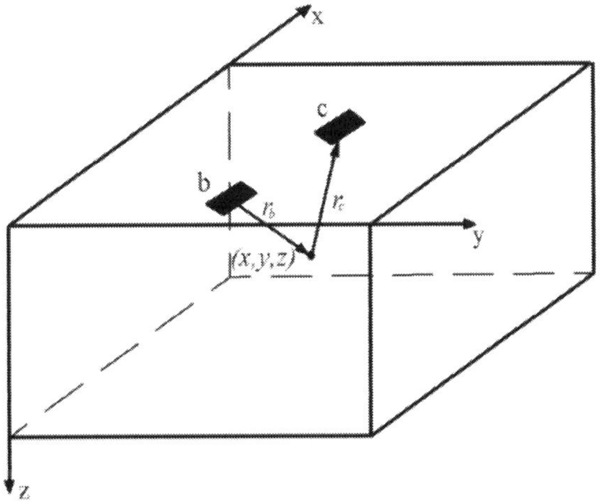

Figure 1. Image reconstruction principles.

$$r_b=\sqrt{(x_b-x)^2+(y_b-y)^2+z^2},\ r_c=\sqrt{(x_c-x)^2+(y_c-y)^2+z^2} \quad (2)$$

In the correlation given above index "b" corresponds to the number of the transmitter and index "c" corresponds to the number of the receiver (see Fig. 1).

The following image reconstruction algorithm is proposed: the controlled area is divided into separate voxels with a given step. For each voxel of the reconstruction area the wave propagation time to and back from the element is calculated. Further the amplitude value of wave propagation time is summed up:

$$A_M(x, y, z)= \sum_{i=1}^{n} A_i(x, y, z) \quad (3)$$

In order to reduce noise in image reconstruction the following correlation is applied:

$$A_M(x, y, z)= \sum_{i\#j} A_i(x, y, z) \cdot A_i(x, y, z) \quad (4)$$

DFA technique enables an accurate determination of the defect location. The results of the inspection are represented as A-scans that display summarized amplitude values of the signals from each separate voxel in accordance with each element of the transducer. This in turn improves the sensitivity and the resolution in the near field of transducer. The resolution of the method depends on the length of the impulse, the size of the phased array and the ultrasound velocity.

The robot-assisted ultrasound tomography system was developed by Tomsk Open Laboratory for Material Inspection (TOLMI) of TPU for the implementation of DFA technique. The system consists of the scanner, the immersion tank, the measurement and control unit. One of the most important parameters that are necessary for the proper inspection is the scanner path. For that reason, it is vital to have the precise model of the controlled object. If the object has a simple geometry, then it is not difficult to generate its model by using CAD system. On the other hand, if the object instead of simple shape has a complex curved shape then modeling can become a difficult, time-consuming procedure.

The study suggests an implementation of optical reconstruction technique that provides a valid way to generate a model of the controlled object [1-3].

3 Experimental Procedure

The composite structure in the shape of a blade was selected for the experiment (see Fig.2). The surface of the sample was not linear and had irregular radial deformations over the whole area.

Figure 2. The object of control.

Artificial defects were performed in the object in order to evaluate the sensitivity of proposed method of control. There were drilled flat-bottom holes with diameters of 3, 6 and 9 millimeters respectively. Each of the defect was placed at a different depth (see Fig. 2).

A very accurate 3-D model of the controlled object surface is required for the correct setting of the scanner path. Three-dimensional reconstruction based on structured light techniques was chosen as approach for model creation. The experimental setup was designed for this purpose (see Fig.3.1). In order to improve the quality of 3-D reconstruction it was suggested to define a sector of the object by covering its surface with matte white coating.

The point cloud in accordance with the controlled object geometry is formed as a result of 3-D reconstruction. Deviation of the results does not exceed 150 micrometers due to the system`s characteristics.

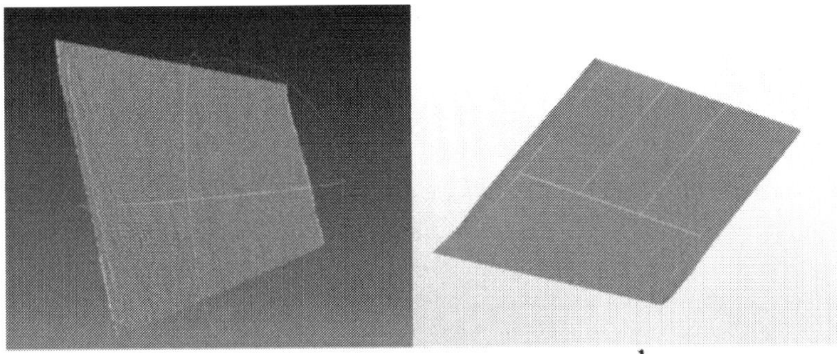

a b

Figure 3. Point cloud – a, surface – b.

The algorithm of scanner path setting allows implementation of T-splines or NURBIS-surfaces. Thus, it is proposed to transform point cloud into NURBIS-surface.

The transformation procedure includes several steps:

1. Transformation of sparse point cloud into regular
2. Creation of the polygonal model by triangulation
3. Generation of the merged grid and its normalization
4. Division of the grid into T-splines
5. Transformation of T-splines into NURBIS-surfaces

Mathematical description of the NURBIS-surface obtainment could be found in [1].

The model of the controlled object was divided into sectors to give more efficient presentation of the results. There are three defects in each of the sectors. The defects are located at the different depth (see Fig.3a and Fig.3b)[4].

The experimental composite sample was scanned by the robotic system of ultrasonic testing. The single channel equipment with 15 MHz transducer was applied.

The first step of the experiment is intercallibration of the scanner and the object of control, that can be performed by optical or tactile probing methods. The scanning path is generated by the means of specialized software in accordance with the surface that was preliminary generated.

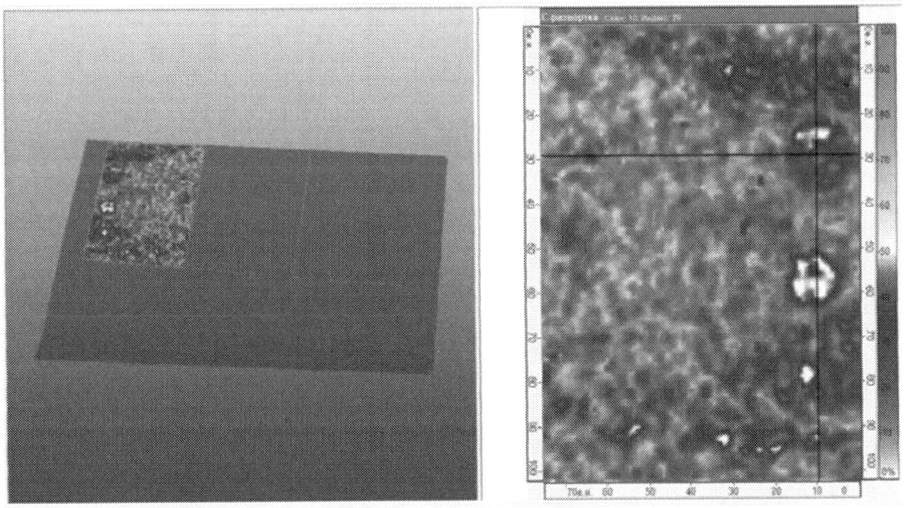

a b

Figure 4. Reconstruction of the first sector-scan (a – on the object, b – C-scan).

Further the scanner is placed into the work position with the surface of the controlled object. Sensitivity setting is arranged. The main parameters that should be set are the following: amplification gain and time corrected gain (TCG).

After that, an automatic procedure starts which enables further analysis of the obtained data.

Additional noise and complex geometry of the bottom surface relief make it rather difficult to reconstruct all the defects at the same image. As noted earlier, this process could be simplified by division the object of control into several sectors. Therefore, the results for the defects located at the same depth are presented as individual sector-scans.

Figure 4a and 4b demonstrates indications of ultrasound scans relevant to the defects with diameters of 3 and 6 millimeters. The defect with diameter of 9 mm is not quite noticeable due to the fact that its depth of occurrence is close to zero. This is caused by the rough bottom surface relief.

The indications of all three defects are defined very clearly at the second sector-scan. The level of the additional noise is rather slight (see Fig.5a and Fig.5b).

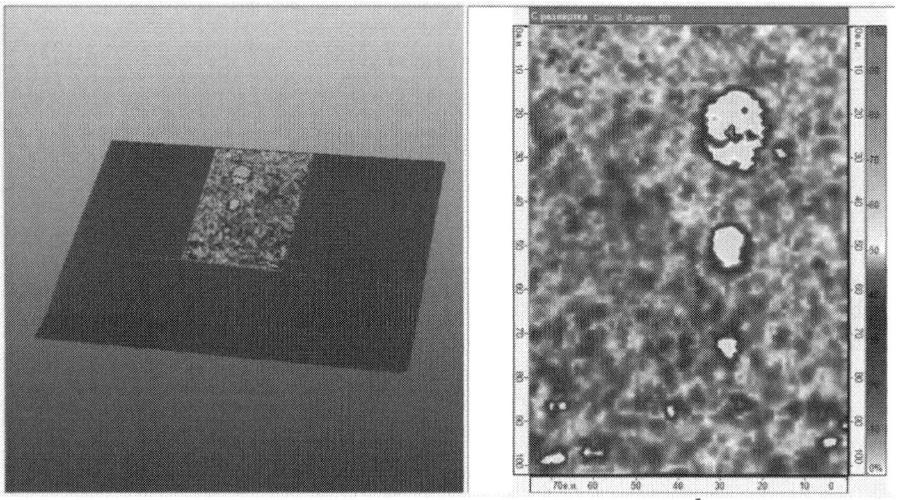

a b

Figure 5. Reconstruction of the second sector-scan (a – on the object, b – C-scan).

The defects of the third sector are located close to the surface and that causes superposition of defect indications and indications of surface reflected signals (see Fig.6a and Fig.6b). However, pulse transmission method allows to get clear indications of defects projections.

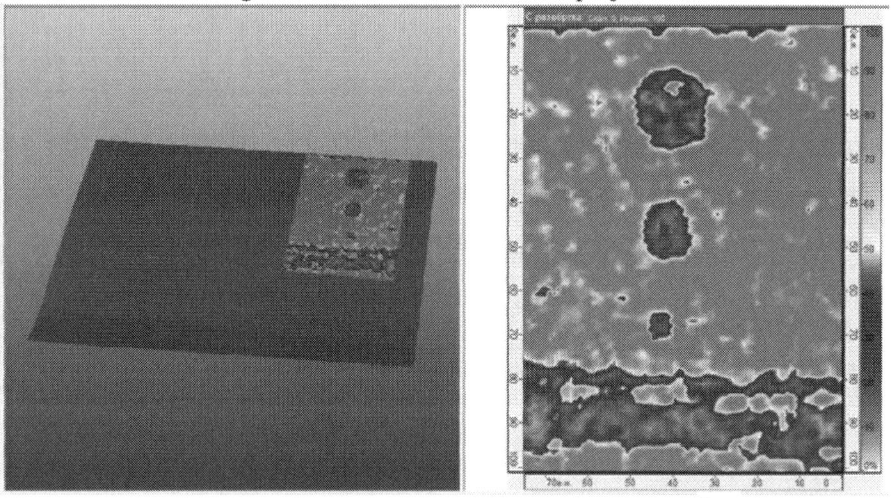

a b

Figure 6. Reconstruction of the third sector-scan (a – on the object, b – C-scan).

4 Conclusion

The accuracy of the inspection results can be affected not only by the parameters of the equipment and its calibration efficiency, but also by the scanner path. The scanner path is to be set exactly in line with the shape of the object to provide positioning of the transducers with multiple axis robotic manipulator. Due to the fact that the scanner path is formed according to the CAD-model of the object, the time that is needed for the model generation is important.

The implementation of optical reconstruction technique allowed to generate CAD-models of controlled objects with an accuracy of 150 μm. The obtained experimental results demonstrate that the proposed method of the surface model reconstruction provides effective nondestructive control of complex shaped carbon fiber composite materials. Moreover, within the results it is possible to identify 8 out of 9 defects indisputably.

Acknowledgement

The research was supported by Russian Governmental program "Science", research project No. 1524.

References

1. I. O. Bolotina, M. E. Djyakina, M. K. Krening, F. Mokhr, K. M. Reddi, A. I. Soldatov, E. Zhantlesov, Russ. J. Nondestruct. **49**, 145 (2013)
2. D. Sednev, A. Lider, D. Demyanuk, M. Kroening, Y. Salchak, Phys.Procedia **70**, 505 (2015)
3. M. Spies, W. Jager, Ultrasonics **41**, 125 (2003)
4. D. Lanman, D. Crispell, G. Taubin, 3DIM **3**, 107 (2007)

MATEC Web of Conferences **48**, 03006 (2016)
DOI: 10.1051/matecconf/20164803006

Small-angle tomography algorithm for transmission inspection of acoustic linear array

Alexey Soldatov[1, a], Andrey Soldatov[1], Masha Kostina[1], Petr Bakanov[2] and Marina Polonskaya[1]

[1] *National Research Tomsk Polytechnic University, 634050 Tomsk, Russia*
[2] *Ultrasonic technology Ltd., 634027 Urozhayniy bystreet, Russia*

Abstract. The paper describes the algorithm of reconstruction of tomographic image used in the through-transition method in a small angle sounding of acoustic linear arrays and the results of practical application of the proposed algorithm. In alternate probing of each element of emitting array and simultaneous reception of all elements of the receiving array is a collection of shadow images of the testing zone. The testing zone is divided into small local areas and using the collection of shadow images computed matrix normalized transmission coefficients for each of the small local area. Tomographic image control zone is obtained by submitting the resulting matrix of normalized transmission coefficients in grayscale or colors.

1 Introduction

Through-transition method has a higher sensitivity and noise immunity, and also has no dead area, in comparison with the pulse-echo method. This method is particularly effective in the testing of the multi-layer composites are widely used at present in aircraft and aerospace [1]. Traditionally, two transducers, emitter and receiver, are used in this method. They are arranged on the opposite sides of the testing object [2, 3]. But, to use only two sensors will not make it possible to determine the distance to the defect [4]. In order to calculate the distance to the defect one must have the projections from different angles, as it is implemented in the computed tomography, which has been widely developed in recent years [5]. It is possible to use two different methods of the acoustic testing: mechanical scanning of the testing object with two sensors - a receiving and emitting, with its position that is fractional to the object testing by discrete and independent of each other, or an electronic scanning of the testing object with two linear arrays.

The advantage of the first option - the possibility of setting an arbitrary step and tracing.
Disadvantages: - long time of the scanning, because of the mechanical sensors travel, - the complicacy and high cost of high-precision mechanical scanner with a programmable control, - the need for high stability of measuring system.

The advantage of the second option – possibility to have a high-speed scanning, but only under such condition – with a fixed geometrical structure of a concrete device. Disadvantage: need of a high stability of equipment when sensor of antenna arrays is emitted.

[a] Corresponding author : asoldatof@mail.ru

2 Problem statement

In practice, the electronic scanning is used. It should be noted that when the acoustic testing is used it is impossible to obtain a collection of projection angles from 0 to 360 degrees. When linear arrays are had the electronic scanning we can have a limited number of projections with a small angular deviation, under the testing of planar structures such as multilayer composites as a rule, no more than ± 30 degrees [6, 7]. Firstly, limitation is linked with the need, in order to provide access from different sides, and secondly, to provide an acoustical coupling sensor with the object of testing. Therefore, it is of interest to study the possibilities of small-angle sensing. For this purpose it is proposed 2D-model of acoustic through-transition detection system with linear arrays and logical and analytical image reconstruction algorithm (visualization) of the internal structure of the object in the testing zone.

3 Two-dimensional model

A two-dimensional model of the acoustic patch with the emitting and receiving arrays is shown in Figure 1.

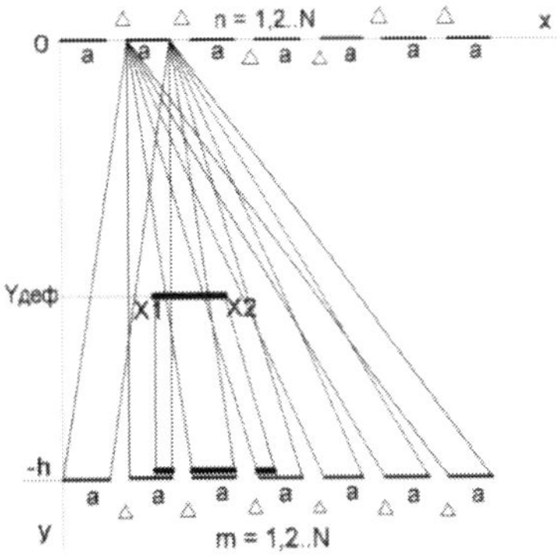

Figure 1. The acoustic path of through-transition method with linear arrays.

Each acoustic arrays consists of N elements with sizes of element "a" equal to the actual diameter of cylindrical piezoelectric element and placed at the same distance Δ from each other. The latter condition is not fundamentally necessary for a real system and introduced only for the convenience of the analytical description of the model. Antenna arrays, dispose on the opposite sides of the testing zone, located collinear with the x-axis at a distance h from one another. Thus, the area of the testing zone is:

$$S = \left[\left(a + \Delta \right) N - \Delta \right] h, \tag{1}$$

where a - the size of the piezoelectric element, Δ - the distance between the piezoelectric element, N - number of elements in the array, h - the distance between the antenna arrays.

Sounding of the testing zone is carried out by means of alternately switching of the emitting sensors. All these elements of the receiver array are recorded by signal that has been passed through the testing object [8]. The result of monitoring procedures is the transmission coefficient matrix of size (N * N). The coefficients are calculated in accordance with the formula:

$$k_{m,n} = A_{m,n} / A_0 , \tag{2}$$

where $A_{m,n}$ – the signal amplitude at the receiver m when the transmitter n shots and the object has the defect, A_0 – the signal amplitude at the receiver m when the transmitter n shots and the object doesn't have the defect.

When the testing object doesn't have the defects then the coefficient matrix of transmission is an unitary matrix. It is assumed that elements of the transmitter and receiver are isotropic in the model, the medium loss equals zero, the radiation from n element of the emitting array into m element of the receiving array is equivalent of geometrical acoustics within the zone (radiation tape), formed by straight lines connecting the beginning of the radiation element to the beginning of the receiver element and the end of the radiation element to the end of the receiving element, and diffraction is ignored. (The concept of radiation tape was introduced for 2D-model and analogical with the meaning of Williams tube (lightpipe) for 3D-model). Zone borders described by the equations of straight lines [9], which are convenient to represent, in accordance with, in the form:

the left (odd) boundary
$$\frac{x - x1(n)}{x2(m) - x1(n)} = \frac{y - y1(n)}{y2(m) - y1(n)}, \tag{3}$$

the right (even) boundary
$$\frac{x - (x1(n) + a)}{(x2(m) + a) - (x1(n) + a)} = \frac{y - y1(n)}{y2(m) - y1(n)}, \tag{4}$$

where $x1\ (n)$ – the left end n-th element of radiating array ($n = 1,2..N$), $x2\ (m)$ – the left end of the m-order element of receiving array ($m = 1,2..N$), $y1\ (n)$ and $y2\ (m)$ – the coordinates of the radiating and receiving elements arrays. They are easy to take equal $y1\ (n) = 0$ and $y2\ (m) = -h$.

A defect in the model is simulated by a straight line segment that is parallel to the x axis, and it is located in the testing zone with the coordinates $X1$, $X2$, Y. The defect supposed to be opaque. The defect attenuation depends on the overlap of radiation tape (Figure 2). Therefore, the transmission ratios are calculated in accordance with formula [10]:

$$k_{m,n} = 1 - S_{m,n}/S_0 \tag{5}$$

where $S_{m,n}$ – the shadow length from the defect on the receiver m for angle of the emitter n, S_0 – the length of the receiver.

If there is not defect in the radiation tape, then $k_{m,n} = 1$. If the defect covers the radiation tape on the half, then $k_{m,n} = 0.5$.

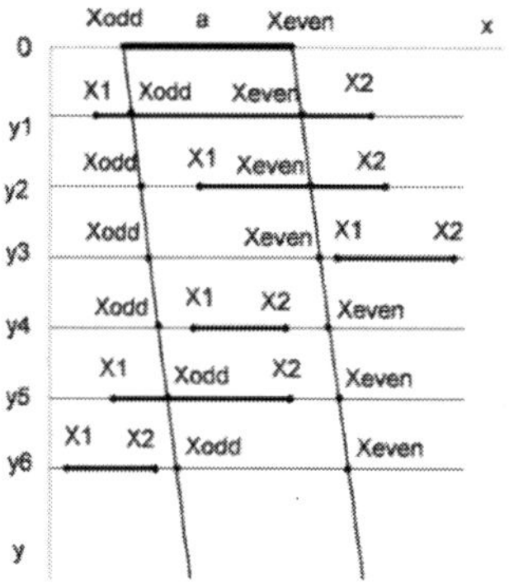

Figure 2. Options of the defects location in the testing zone, $X1$, $X2$ – coordinates of the defect boundaries, x_{odd}, x_{even} – coordinates of the radiation boundaries tape at different level Ydef (Y1, Y2 and etc.).

Options of the possible defect location relatively to the radiation tape and the corresponding transmission coefficients are summarized in Table 1.

Table 1. Options of the possible defect location.

Conditions characterizing the defect placement in the testing zone	transmission factor $k_{m,n}$
$X1 \le x_{odd}(n,m) < x_{even}(n,m) \le X2$	0
$x_{odd}(n,m) < X1 < x_{even}(n,m) \le X2$	$0 < X1 - x_{odd}(n,m)/a < 1$
$x_{odd}(n,m) < x_{even}(n,m) \le X1 < X2$	1
$x_{odd}(n,m) < X1 < X2 < x_{even}(n,m)$	$0 < a - (X2 - X1)/a < 1$
$X1 < x_{odd}(n,m) < X2 < x_{even}(n,m)$	$0 < x_{even}(n,m) - X2/a < 1$
$X1 < X2 \le x_{odd}(n,m) < x_{even}(n,m)$	1

A matrix of $N*N$ transmission coefficients $k_{m,n}$ (Figure 3) was given as an example of simulation results. The results were obtained with linear radiating and receiving arrays located at distance $h = 60$ mm, each array consists of $N = 8$ elements with size $a = 6$ mm and distance between elements is $\Delta = 3$mm. Defect coordinates are $X1 = 32.1$ mm, $X2 = 37.1$ mm and $Y_{def} = -40$ mm.

$$A = \begin{pmatrix} 1 & 1 & 1 & 1 & 1 & 1 & 0.85 & 0.167 \\ 1 & 1 & 1 & 1 & 1 & 0.85 & 0.167 & 0.817 \\ 1 & 1 & 1 & 1 & 0.85 & 0.167 & 0.817 & 1 \\ 1 & 1 & 1 & 0.85 & 0.167 & 0.817 & 1 & 1 \\ 1 & 1 & 0.85 & 0.167 & 0.817 & 1 & 1 & 1 \\ 1 & 0.85 & 0.167 & 0.817 & 1 & 1 & 1 & 1 \\ 0.85 & 0.167 & 0.817 & 1 & 1 & 1 & 1 & 1 \\ 0.167 & 0.817 & 1 & 1 & 1 & 1 & 1 & 1 \end{pmatrix}$$

Figure 3. Matrix of coefficients $k_{m,n}$, when the location of defect coordinates are $X1 = 32.1$ mm, $X2 = 37.1$ mm and $Y_{def} = -40$ mm.

A graphical matrix representation of transmission coefficients $k_{m,n}$ in grayscale is shown in figure 4.

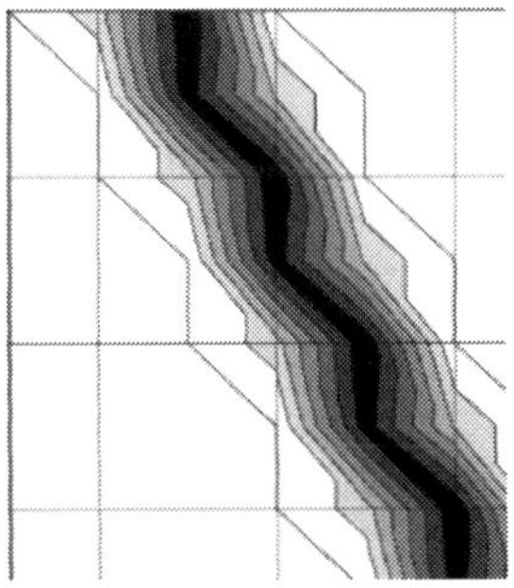

Figure 4. A graphical matrix representation of the coefficients $k_{m,n}$.

Analysis of a graphical matrix representation of the transmission coefficients $k_{m,n}$ is shown, if the testing zone has a defect then it is possible:
- determine the position of the defect on the slope of the line, representing a linear approximation of the location of the minima of the transmission coefficients $k_{m,n}$ in the matrix rows, for example, performed by the least square method,
- make an assessment of the size of the defect and its transfer properties by the transmission coefficients value $k_{m,n}$,
- it can be inferred by the linear dimensions on the bandwidth of the minima defect of the matrix coefficients $k_{m,n}$ in the graph.

4 Calculation of the matrix transmission coefficients

The greatest interest is the investigation of the possibility to evaluate the state of the testing object at each point of the testing zone when one use the transmission matrix coefficients is obtained, in this case, with linear acoustic arrays that is equivalent to obtain the tomographic image of the testing zone. The matrix transmission coefficients for the model with acoustic linear arrays from N elements each represents ($2N$-1) truncated collections of radiation projection, transmitted through the testing zone to the receiving array at angles from 0° to $\pm \varphi\ (k)$:

$$\varphi(k) = arctg\left(k(a+\Delta)/h\right), \tag{6}$$

where $k = 0,1..(N\text{-}1)$.

Moreover, if collections of projections for $k = 0$ is equal to N (the only complete), for $k = 1$ is equal to (N-1) and so on, then for $k = (N\text{-}1)$, it is equal only to one (unitary) [11].

It is impossible to use Radon's transformation [12] for visualization because we have a truncated collections of projections from emitters to receivers that probe the testing zone at different angles. In this regard, for the reconstruction of tomographic image of the testing zone the algoritm was proposed and its modeling was carried out in Mathcad .The algorithm is based on the following principles:

- testing zone is divided into small local areas that are considered to be point,
- every point of space in the testing zone (between the emitting and receiving arrays) because of its abnormality (defects) has the affect on the transmission coefficients $k_{m,n}$. Thus each specific point affects to the value $k_{m,n}$ of those radiation tapes, where the point comes to these fields,
- the resulting attenuation characterized by abnormal of this concrete point as it was supposed to define as the sum of the transmission coefficients $k_{x,y} = \sum_{1}^{P(x,y)} k_{m,n} / P(x,y)$, divided into the number of radiation tapes that were passed through this point (normalized transfer coefficient for a point with coordinates x and y),
- if even one or more of the coefficient $k_{m,n}$ of radiation tapes, in which the given point came, occurs equal to one (transmission in the radiation tape is maximum in this case), then the resulting normalized sums of $k_{x,y}$, for the given point, it should be taken up as equal to one, because this point of space does not inject the losses (correcting condition).

The structure of tomography reconstruction algorithm is shown in figure 5.

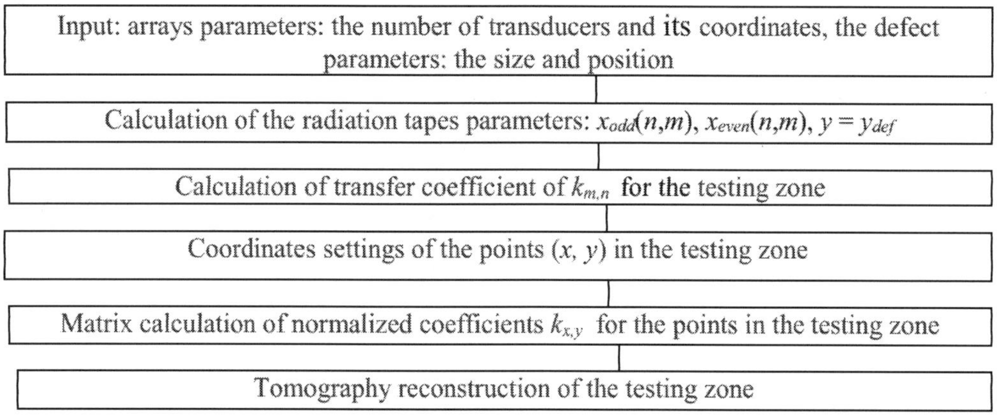

Figure 5. Tomography reconstruction algorithm.

The calculation result of the matrix transmission coefficients $k_{x,y}$ in the testing zone in the proposed algorithm on the dates analogical with figure 4, shown in figure 6. The point of space in the testing zone were set at a distance of 2.5 mm from one another in the coordinates of «x» and «y».

D =

0.979	1	1	1	1	1	1	1	1	1	1	1	1	1	1	1	1	1	1	1	1	1
0.877	0.979	1	1	1	1	1	1	1	1	1	1	1	1	1	1	1	1	1	1	1	0.979
0.754	0.836	0.97	1	1	1	1	1	1	1	1	1	1	1	1	1	1	1	1	1	0.97	0.836
1	0.803	0.836	0.975	1	1	1	1	1	1	1	1	1	1	1	1	1	1	1	0.975	0.836	0.803
0.854	0.975	0.86	0.86	0.975	1	1	1	1	1	1	1	1	1	1	1	1	1	0.975	0.86	0.86	0.975
0.833	0.833	0.836	0.858	0.877	0.979	1	1	1	1	1	1	1	1	1	1	1	0.979	0.877	0.858	0.836	0.833
1	0.767	0.806	0.836	0.836	0.836	0.979	1	1	1	1	1	1	1	1	1	0.979	0.836	0.836	0.836	0.806	0.767
0.854	0.97	0.831	0.833	0.86	0.803	0.858	0.979	1	1	1	1	1	1	1	0.979	0.858	0.803	0.86	0.833	0.831	0.97
0.854	0.854	0.806	0.75	0.812	0.835	0.773	0.858	0.967	0.981	1	1	1	0.981	0.967	0.858	0.773	0.835	0.812	0.75	0.806	0.854
0.661	0.767	0.833	0.806	0.806	0.75	0.835	0.754	0.857	0.963	0.981	1	0.981	0.963	0.857	0.754	0.835	0.75	0.806	0.806	0.833	0.767
0.803	0.97	0.944	0.833	0.812	0.813	0.731	0.731	0.735	0.841	0.925	0.919	0.925	0.841	0.735	0.731	0.731	0.813	0.812	0.833	0.944	0.97
0.854	0.854	0.767	0.831	0.806	0.75	0.727	0.724	0.713	0.675	0.728	0.85	0.728	0.675	0.713	0.724	0.727	0.75	0.806	0.831	0.767	0.854
0.806	0.767	0.806	0.806	0.812	0.813	0.813	0.708	0.69	0.543	0.47	0.486	0.47	0.543	0.69	0.708	0.813	0.813	0.812	0.806	0.806	0.767
1	0.97	0.833	0.833	0.806	0.75	0.731	0.708	0.65	0.535	0.365	0.167	0.365	0.535	0.65	0.708	0.731	0.75	0.806	0.833	0.833	0.97
0.854	0.833	0.833	0.854	0.812	0.813	0.727	0.708	0.665	0.6	0.557	0.47	0.557	0.6	0.665	0.708	0.727	0.813	0.812	0.854	0.833	0.833
0.854	0.806	0.831	0.806	0.75	0.767	0.813	0.727	0.642	0.733	0.785	0.817	0.785	0.733	0.642	0.727	0.813	0.767	0.75	0.806	0.831	0.806
1	0.97	0.836	0.833	0.806	0.767	0.75	0.746	0.746	0.846	0.939	1	0.939	0.846	0.746	0.746	0.75	0.767	0.806	0.833	0.836	0.97
0.806	0.806	0.833	0.835	0.75	0.831	0.746	0.815	0.867	0.959	1	1	1	0.959	0.867	0.815	0.746	0.831	0.75	0.835	0.833	0.806
0.854	0.833	0.833	0.806	0.833	0.769	0.829	0.85	0.977	1	1	1	1	1	0.977	0.85	0.829	0.769	0.833	0.806	0.833	0.833
0.963	0.97	0.806	0.833	0.736	0.85	0.855	0.974	1	1	1	1	1	1	1	0.974	0.855	0.85	0.736	0.833	0.806	0.97
0.611	0.806	0.806	0.769	0.85	0.873	0.974	1	1	1	1	1	1	1	1	1	0.974	0.873	0.85	0.769	0.806	0.806
0.854	0.854	0.963	0.831	0.831	0.974	1	1	1	1	1	1	1	1	1	1	1	0.974	0.831	0.831	0.963	0.854
0.854	1	0.855	0.855	0.974	1	1	1	1	1	1	1	1	1	1	1	1	1	0.974	0.855	0.855	1
1	0.831	0.855	0.974	1	1	1	1	1	1	1	1	1	1	1	1	1	1	1	0.974	0.855	0.831
0.831	0.855	1	1	1	1	1	1	1	1	1	1	1	1	1	1	1	1	1	1	1	0.855
0.873	1	1	1	1	1	1	1	1	1	1	1	1	1	1	1	1	1	1	1	1	1

Figure 6. Matrix of the distribution of the resulting transfer coefficients of $k_{x,y}$ in the testing zone with the location of the defect with coordinates $X1$ = 32.1 mm, $X2$ = 37.1mm, Ydef = -40 mm.

5 Results

If each value in the normalized coefficient matrix to represent as grayscales, we will obtain a tomographic image of the testing zone (figure 7).

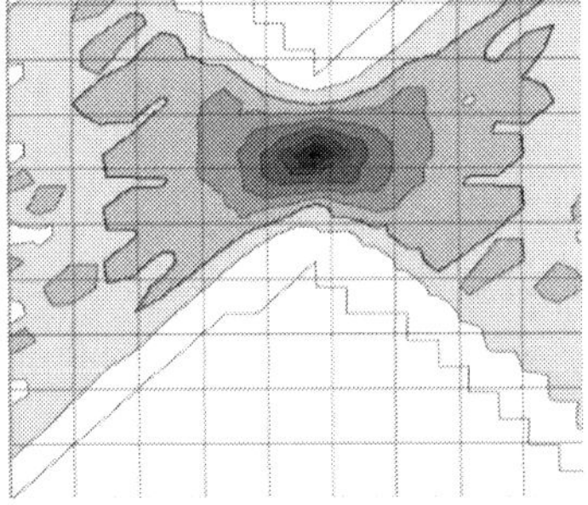

Figure 7. Tomographic image of the testing zone, corresponds to the normalized coefficient matrix.

Result shown in figure 7 was obtained without using the correction of the conditions: if even one or more of the coefficient $k_{m,n}$ of radiation tapes, which belongs to this point, it is equal to one, then $k_{x,y}$, passed through this point is equal to one.

Figure 8 shows the imaging testing zone with the defect with coordinates $X1$ = 17,1mm, $X2$ = 25,1 mm and ydef = -30 mm without using the conditions correction (a) and with its using (b) for N = 4, a = 6 mm, Δ = 3 mm, h = 60 mm, the distance between the points in the testing zone is 5 mm with total number of points M = 66 (100%). Thus, if in the first case we record as the abnormal points 53 (80.3%), while in the second case, only 18 (27.7%), this increase the localization of the defect position significantly. However, the area of the minimum values of the coefficient $k_{x,y}$ remained unchanged.

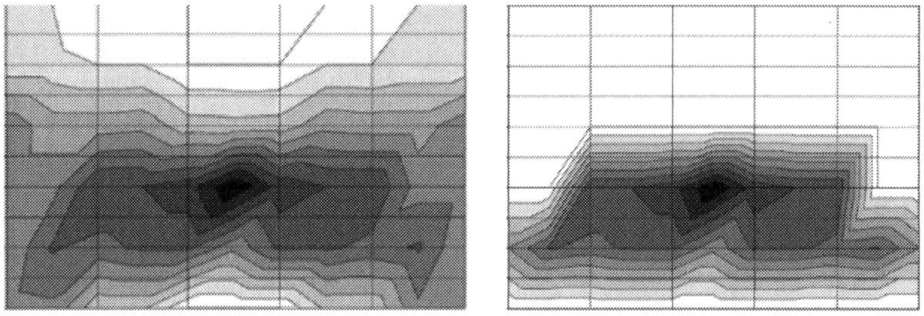

a b

Figure 8. Tomographic image of testing zone without correction (a), and with correction (b), for N = 4, a = 6 mm, Δ = 3 mm, h = 60 mm, the coordinates of the defect $X1$ = 17,1 mm, $X2$= 25,1 mm and $Ydef$ = -30 mm.

6 Conclusion

The performed simulations have shown the possibility of tomographic reconstruction of the testing zone at a small-angle sensing with linear acoustic arrays use. The algorithm allows you to specify the number of local points in the testing zone, the parameters of acoustic arrays: the number of emitting and receiving elements, the distance between them, as well as the distance between the arrays. Application of the proposed algorithm of tomography reconstruction of the testing zone allows to localize the spatial position of the defect with an error of no more than 10% of the size of the testing zone. The change of the local areas number in the testing zone, with the fixed parameters of the acoustic path, does not significantly increase the accuracy of the localization of the spatial position of the defect. It is possible to increase the accuracy by means of increasing the number of angles and angular deflection angles that is achieved by means of increasing the number of elements in the arrays.

References

1. A. Tovstik, A. Zakarluka, A. Petrenko, Herald of the Dnepropetrovsk University **17**, 96 (2014)
2. R. Rausutis, A. Voleisis, R. Kazys, Ultragarsas **63(3)**, 15 (2008)
3. P. He, IEEE Trans. Ultrason., Fer., Freq. Contr. **45(1)**, 114 (1998)
4. F. Reed, T. Batzinger, R. Reed, Rev. of Pr. in QNE **12**, 1265 (1993)
5. G. Herman, Fundamentals of computerized tomography: Image reconstruction from projection (Springer-Verlag London, 2009)
6. R. Rausutis, R. Kazys, L. Mazeika, NDT&I **40**, 324 (2007)
7. R. Rausutis, R. Kazys, L. Mazeika, NDT&I **41**, 300 (2008)
8. K. Kvasnikov, A. Soldatov, I. Bolotina, M. Krening, A. Potapenko, Rus. J. NDT **49(11)**, 625 (2013)
9. E. Magrab, *An Engineer's Guide to Mathematica* (John Wiley & Sons, 2014)
10. D. Buy, V. Makarov, Al. Soldatov, An. Soldatov, Intern. J. A&B. Res. **12**, 216 (2015)
11. A. Goncharsky, S. Romanov, S. Seryozhnikov, WM **51(3)** 389 (2014)
12. S. Deans, *The Radon Transform and Some of Its Applications* (DP, N.Y. 31 East 2-nd str., Mineola, 2007)

MATEC Web of Conferences **48**, 03007 (2016)
DOI: 10.1051/matecconf/20164803007

Numerical simulation of the betatron magnetic field using ELCUT software

Ivan Zatonov[1,a] and Mikhail Shtein[1]

[1] *National Research Tomsk Polytechnic University, 634050 Tomsk, Russia*

Abstract. The article examines the possibility of using software kits to solve multi-physical tasks in betatron engineering. It emphasizes radiation control in aerospace industry using small-size pulse betatron as a source of ionizing radiation. Main characteristics of the betatron magnetic field are calculated using Elcut software kit. Dependence of calculation accuracy on the grid quantity is studied. Focusing properties of the betatron magnetic field are analyzed.

1 Introduction

Aerospace industry plays a significant role in ensuring military security, scientific and economic development of the Russian Federation. At this stage of development, Russia moves from sustaining the previous level of aerospace industry to new ambitious projects aiming at a higher production potential of this area [1]. Such an important role in the Russian strategy of military and technical and economic development demands a high level of reliability from the aerospace industry, which depends, first of all, on using instruments and methods of non-destructive testing at all production stages.

Owing to its increased feedback, radiation control is widely used for inspection of aerospace industry related products. The point of the digital X-ray method is the imaging of the object structure identifying foreign material and its geometry in the structure of the control object. In the last few years, similar techniques have been used to solve tasks of the aviation security. Inspection of large-size articles and air containers requires high energy X-ray sources now-a-days represented by betatrons and linear accelerators [2]. The distinctive feature of betatrons is small size and low cost, which allows them competing on the market regardless of the low dose rate.

The first Russian betatron for the energy of 4 MeV was launched in Tomsk Polytechnic University in 1948. Since then, TPU is the reference University for betatron engineering not only in Russia but worldwide.

[a] Corresponding author : ivan.zatonov@yandex.ru

2 Experimental procedure

2.1 Betatron setting

During betatron manufacturing, there are usually difficulties related to its adjustment, i.e. finding magnetic field parameters to achieve the rated electron energy and maximum dose rate.

There are several ways to adjust betatrons:

- Adjustment by searching parts that contribute into distribution of the betatron magnetic field and measurement of the dose rate for each variant. To ensure correct adjustment and maximum possible dose rate, dozens or hundreds of combinations are required sometimes.
- Magnetic measurements and change of magnetic field parameters in inter-polar gap based on the results of measurements. This brings a good result in a few combinations.

The principle of betatron operation is based on the fact that an electron can move along a circular orbit of the constant radius gaining energy in the magnetic field that decreases in radius and increases in time. Energy increment over one turn is proportional to the speed of magnetic flux change inside the orbit. Any deviation of the electron from the constant orbit results in bringing it back to this orbit by magnetic forces. That is why its radius r_0 is called equilibrium.

One of the main tasks of adjustment is to find the optimum radius of the equilibrium electron orbit as this radius determines both accelerated electron energy, and area of magnetic focusing forces, hence their quantity.

It is possible to find the equilibrium radius using a simple condition called "two to one relation" in the betatron theory [3].

$$2B_z(r) = \overline{B}, \quad (1)$$

where $B_z(r)$ is the induction of the magnetic field on the equilibrium radius r, $\overline{B}$ is the average induction in the circle r_0.

The same values that are found through magnetic measurements determine the energy of accelerated electrons

$$E = \sqrt{(300 r_0 B_z(r_0))^2 + \mathrm{E}_0^2} - \mathrm{E}_0, \quad (2)$$

where E_0 - is the electron rest energy, which is 0,511 MeV.

The results of magnetic measurements are used to determine the radius dependence of the magnetic field decay index [4].

$$n = -\frac{\mathrm{d}(B_z(r))r}{B_z(r)\mathrm{d}r}. \quad (3)$$

The potential function of magnetic field tells better about its focusing properties [5]

$$V_\mu = \frac{\mathrm{e}A^2}{2\mathrm{mc}^2} = \frac{\mathrm{e}}{2\mathrm{mc}^2}\left(\frac{r\overline{B}}{2}\right)^2, \quad (4)$$

where e - is the electron charge, A - is the vector magnetic potential, m -is the electron mass, c - is the speed of light.

2.2 Numerical simulation

The task of this work was to demonstrate the possibility to achieve the magnetic field characteristics for actual betatron designs with the required accuracy using numerical simulation as compared to time-consuming magnetic measurements. Numerical simulation allows significantly reduce the time of betatron adjustment.

Such a simulation can be done using software kits designed for finite element analysis in different areas of physics and engineering, including multi-physical tasks. The currently used kits are Ansys, Comsol Multiphysics, Elcut and so on.

Software kit Elcut was used for numerical simulation of the main characteristics of the betatron magnetic field as being most simple and easily understood. The design of a small-size MIB-10 betatron was used for analysis.

The geometry of the betatron magnet is shown in Figure 1.

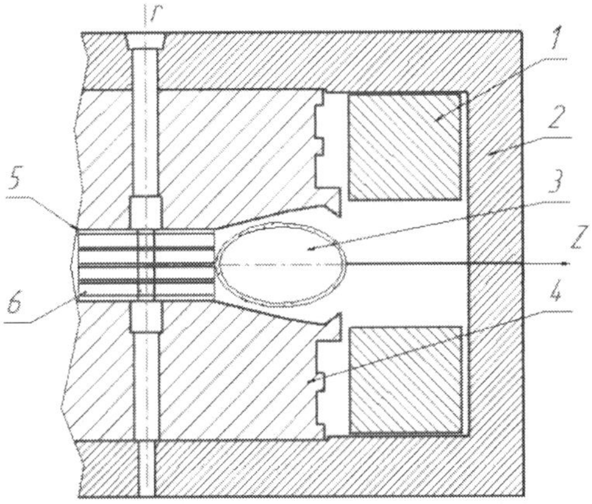

Figure 1. Geometry of a betatron, symmetric to r axis, where 1- coil with 7500 ampere-turns current; 2- reverse magnetic conductor; 3- betatron X-ray chamber; 4- magnetic pole pieces; 5- five identical magnetic spacers; each 2.4 mm thick; 6- four magnetic pies; each 8 mm thick.

With account for the axially symmetric design of the MIB-10 betatron, a geometric model was made in Elcut software kit followed by setting the physical parameters of objects. The work with geometric model is done in a cylindrical coordinate system in half-plane zr ($r \geq 0$) (Figure 2).

Since the finite-element method is used to find the solution in the software, the grid application is a mandatory stage of modelling. The smaller the grid space, the higher the accuracy of calculations but the longer the time. Initially a number of calculations was done, the spaces being 2, 1, 0.5 and 0.25 mm. The results do not change beginning from 1mm space. Setting an option of automatic selection of the grid space with 13403 grid points had the same result. The grid at automatic selection is shown in Figure 2b.

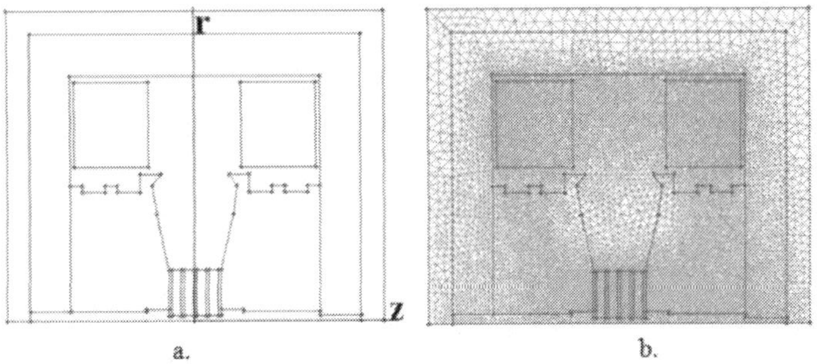

Figure 2. a. Geometric model of a betatron in axially symmetric class of the model; b. Automatic sampling of the grid applied to the studied betatron model.

The Elcut software solves the Poincare equation for vector magnetic potential A but in the axially symmetric task it calculates the related flux function $\Phi = rA$. Then vector components of the magnetic induction are found by numerical differentiation B_z and B_r.

First, the value of the radius of equilibrium orbit was determined based on assumption that the orbit is completely within the z plane. For this, the contour at z=0 mm on sector $r \in [43;130]$ mm was studied by the software, and induction values depending on radius $B_z(r)$ were found.

The average induction B_{cp} in not calculated by the software; it was found as follows:

$$\overline{B}(r) = \frac{\Phi(r)}{S} = \frac{\int d\Phi}{S} = 2\frac{\int_0^r B_z(r)\pi r dr}{\pi r^2}, \tag{5}$$

where $\Phi(r)$ - is the magnetic flux of radius r, S - is the area of this circle.

Since data are presented as a table, the Simpson formula was used for calculation of the certain integral:

$$\overline{B}(r_i) = \frac{\sum_{i=1}^{n} \pi(r_{i+1}^2 - r_i^2)\frac{B_z(r_{i+1}) + B_z(r_i)}{2}}{\pi r_i^2}. \tag{6}$$

The calculated values of the average induction $\overline{B}$ correspond to the average radius:

$$\overline{r_i} = \frac{r_i + r_{i+1}}{2}. \tag{7}$$

3 Results

The cross point of the chart of induction distribution B_z c $\frac{\overline{B}}{2}$ is the value of the equilibrium radius (Figure 3). With the above-mentioned sizes of the pie stack, the value of the equilibrium radius is 101,8 mm, the induction on it being 0,278 Tl.

The value of the electron energy in the given circuit was determined in parallel to the equilibrium radius. Formula (2) was used for calculation and 8 MeV value was obtained, which was close to the expected value.

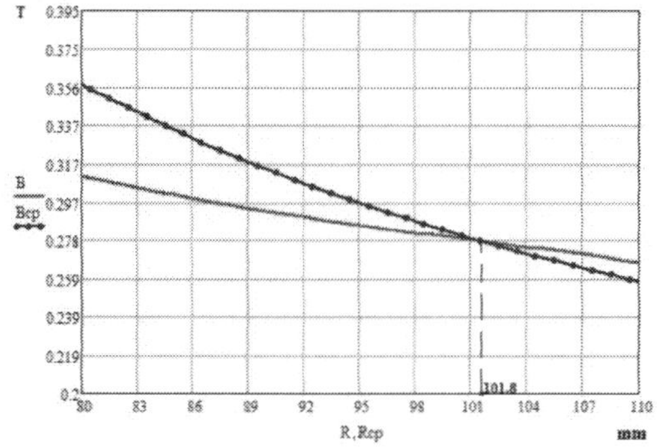

Figure 3. Determining the equilibrium radius.

The indicator of magnetic field strength decay in z=0 plane calculated by formula (3) is given in figure 4.

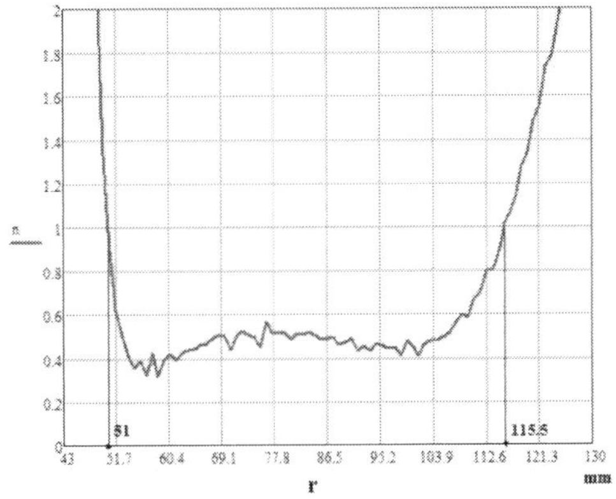

Figure 4. Radius r dependence of the decay indicator n.

Presence of focusing forces in radial and axial directions at the same time is guaranteed when $0<n<1$ inequality is satisfied. According to the chart, the stable orbits are those with radii from 51 to 115,5 mm.

As already said, the potential function V_{μ} is a visual and complete characteristic that describes focusing properties of the magnetic field. The surface described by the potential function is normally called a potential hole [6].

The potential hole made for z=0 plane is shown in Figure 5.

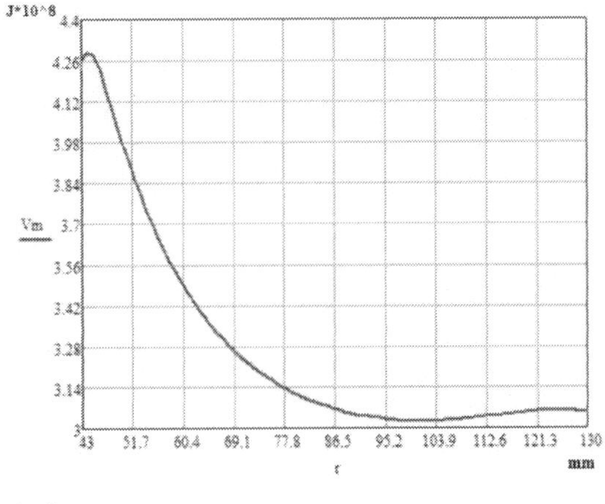

Figure 5. Potential hole for z=0 plane.

As it appears from the betatron theory, the minimum of the potential hole is on the equilibrium radius.

To have a comprehensive idea of focusing forces along the whole cross-section of inter-gap space including along the cross-section of the accelerating chamber, the potential hole is calculated as a function of two coordinates, and usually presented as lines of the same level on the plane. The outline in Figure 6 shows an area, where values of the potential function were calculated, resulting in the potential hole. This picture also demonstrates distribution of the magnetic field lines.

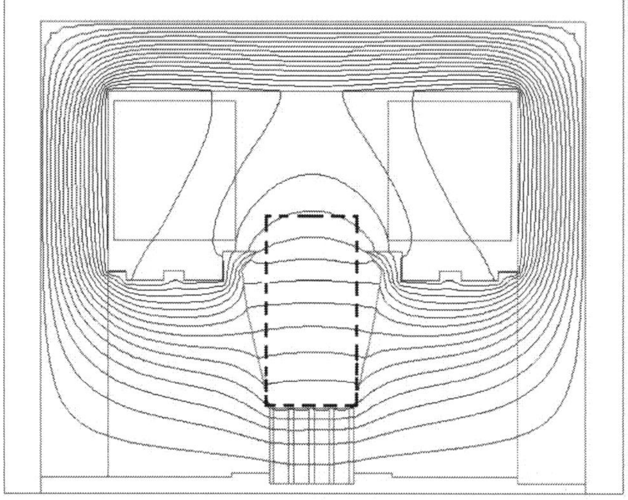

Figure 6. Area of the model used for calculation of potential function.

For the specified conditions of calculations, we get a picture of the potential hole presented in Fig. 7. The red outline shows a vacuum chamber that in theory can have any electron orbits inside of it. The green contour indicates the chamber injector. The latter reduces the effective chamber volume but a betatron cannot work without it.

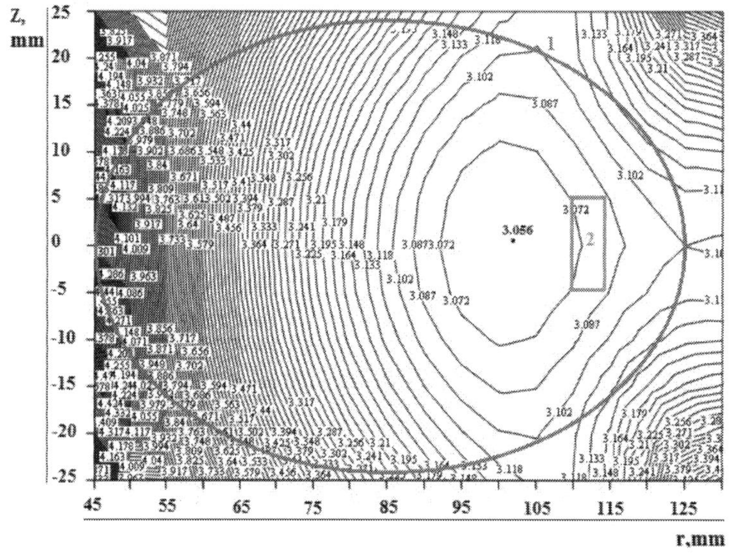

Figure 7. Chart of potential function lines: 1 – inscribed X-ray chamber; 2 – injector.

4 Conclusion

Elcut software suite intended for solving multiphysical tasks by the finite-element method is an effective instrument for numerical modelling of the betatron magnetic field. In many cases, it can replace the process of magnetic measurements and reduce the time of betatron adjustment. Computer time spent for the modelling was not long and took 30 seconds. Processing of the results takes much longer. In this respect, the problem number one is to import the result data of Elcut into more powerful modern programs of data store processing, and create a command file for complete automatization of the process of getting the results in a convenient and easy-to-use form.

As for specific results of the presented magnet modelling, it is obvious that the size of the pie stack was incorrectly selected. The equilibrium radius is close to the injector and only a small part of the chamber cross-section is effective. The radius is set off outwards and the first line of the equal level is in contact with the injector (Fig. 7). Only within this line stable electron orbits may exist. Electrons whose oscillating energy is higher than potential energy of the first level, can make several turns but ultimately they will overcome this barrier and hit the injector.

In order to reduce the equilibrium radius, it is necessary to decrease the height of the pies and increase thickness of non-magnetic spacers leaving the gap between the pole pieces unchanged. By doing several numerical simulations, one can determine the acceptable position of the equilibrium radius and hence the structure of the pie stack.

References

1. A. S. Karpov, Natsionalnyie interesyi: prioritetyi i bezopasnost **12**, 43 (2008)
2. J. Accardo, M. Ahmad Chaudhry, Journal of Radiations Research and Applied Sciences **7**, 198 (2014)
3. K. S. Grishin, Izvestiya TPI **87**, 95 (1957)
4. Timothy W. Koeth, Physics Procedia **66**, 622 (2014)
5. B. N. Rodimov, Izvestiya TPI **87**, 3 (1957)
6. D. A. Noskov, Izvestiya TPI **87**, 151 (1957)

MATEC Web of Conferences 48, 04001 (2016)
DOI: 10.1051/matecconf/20164804001

A study of hardware implementations of the CRC computation algorithms

Evgeniy Mytsko[1,a], Andrey Malchukov[1], Svetlana Ryzova[1] and Valeriy Kim[1]

[1]*National Research Tomsk Polytechnic University, 634050 Tomsk, Russia*

Abstract. The paper is about hardware implementations of the CRC computation algorithms. Combinational circuits of CRC8 and CRC32 computation devices, which can be embedded in satellites for error checking in configuration memory and data transmission module, were considered. The conclusion about the advantages of matrix-driven algorithm hardware implementation, which are a simple diagram is built using only logic «exclusive OR» was done. The examples of CRC8 and CRC32 devices working were presented according to parametric model with different input data.

1 Introduction

For the aerospace and defense electronics, telecommunications and control systems are very important to be able to make sure that the configuration data in the Field-Programmable Gate Arrays (FPGAs) on board does not contain any errors. Ionizing radiation in space can cause unwanted switching of memory cells in the FPGA, which contains the configuration memory, user memory and registers. Such switching may cause equipment failure, which is critical in outer space. CRC (Cyclic redundancy code) is algorithm of checksum computation, which is applied in different data transmission standards, compress algorithms, coding standards and bitmaps [2, 3]. CRC checksums can be used to detect errors and failures in the configuration memory and data transmission modules There are different types of CRC such as CRC8, CRC16, CRC32, which are differed by generator polynomial length [1] and accordingly by the checksum length. In the papers [4, 5] software implementation of table-driven and matrix-driven algorithms were described. The advantages of matrix-driven algorithm implementation, which are concluded in the low requirements of memory, were showed. The asynchronous hardware implementations of CRC8 and CRC32 for FPGA, which can be embedded in satellites, were offered in the paper. The implementation allows to simplify device circuit and to apply features of matrix-driven algorithm in hardware implementation.

2 Hardware implementations of the CRC computation algorithms

2.1 Hardware implementation of the table-driven algorithm

[a] Corresponding author : evgenvt@tpu.ru

Functional diagram of the asynchronous hardware implementation was developed using block-based design in Quartus II for FPGA Cyclone. Figure 1 shows functional diagram of CRC8 computation device. A byte of data which is summed by modulo 2 with *init* [7..0] according to parametric model is supplied to *data* [7..0] input. A byte of CRC is supplied to *crc* [7..0] input.

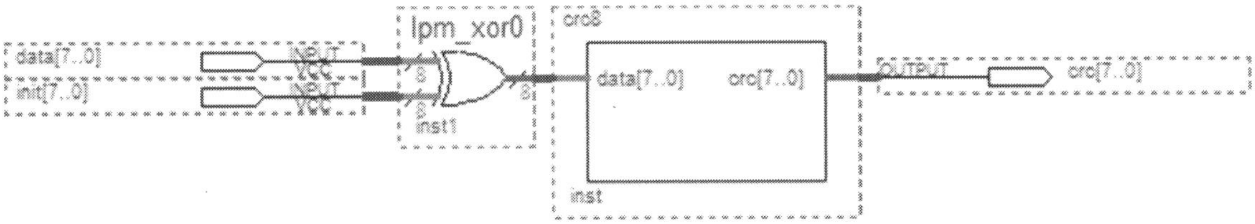

Figure 1. Functional diagram of CRC8 computation device.

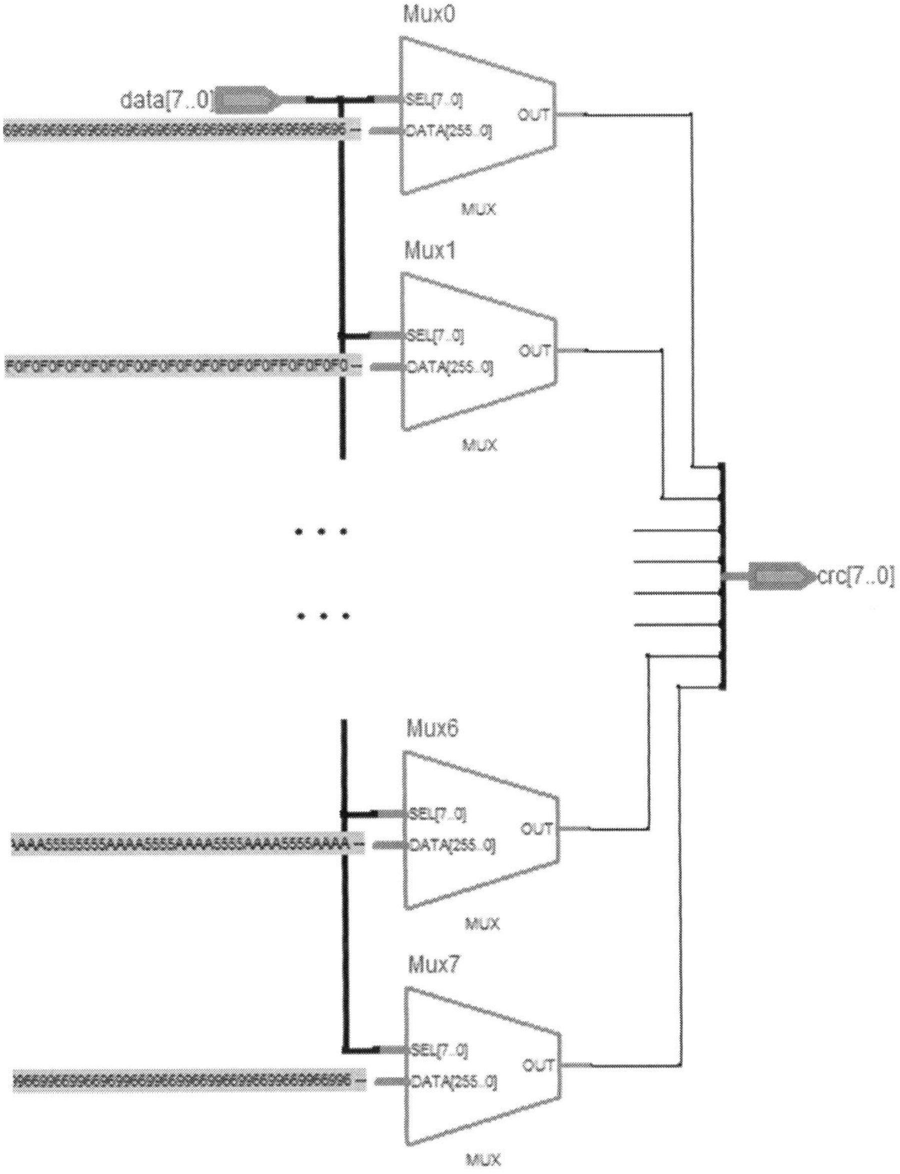

Figure 2. The combinational circuit of hardware implementation of CRC8 table-driven algorithm.

Figure 2 presents the combinational circuit of hardware implementation of CRC8 table-driven algorithm with storage elements (for constant storing) and multiplexers.

According to the table-driven algorithm of CRC computation [4] a byte of data is summed by modulo 2 with checksum byte (the initial value of checksum is chosen by parametric model of algorithm [4]). After that the result of addition is used to index of the table which is filled by reminders of all byte combinations by generator polynomial. The value from the table is a CRC8 checksum. CRC32 computation algorithm is similar to CRC8 with the exception of some features. In this case the length of CRC, the length of generator polynomial and the length of table elements increases to 32 bits.

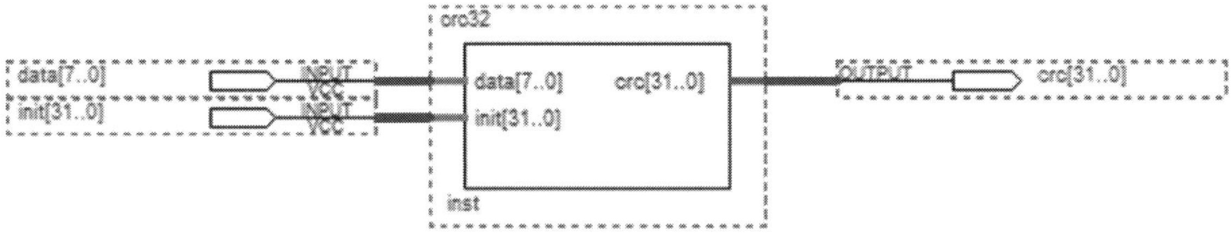

Figure 3. Functional diagram of CRC32 computation device.

For the hardware implementation of CRC32 an additional input *init* [31..0] for circuit is required. 32-bit initial value of CRC32 is supplied to the *init* input according to parametric model. A Low byte of *init* (7 – 0 bits) is summed by modulo 2 with data byte. Then similarly to CRC8 the element of the table is selected and summed by modulo 2 with high three bytes of *init* (31 – 8 bits), which are shifted to the right 8 times. Figure 3 shows functional diagram of CRC32 computation device.

In the CRC32 implementation for table-driven algorithm logic «exclusive OR» will be added to combination circuit and the length of constant will be increased. Herewith the hardware implementation of matrix-driven algorithm of CRC computation allows to avoid multiplexors, storage usage and build combinational circuit using only logic «exclusive OR».

2.2 Hardware implementation of the matrix-driven algorithm

Feature of matrix-driven algorithm of CRC computation is multiplication by modulo 2 of vector (input byte) by matrix [6] which is calculated based on generator polynomial. The feature allows implement matrix-driven algorithm by combinational circuit without storage using only logic «exclusive OR». Figure 4 represents multiplication by modulo 2 of vector by matrix.

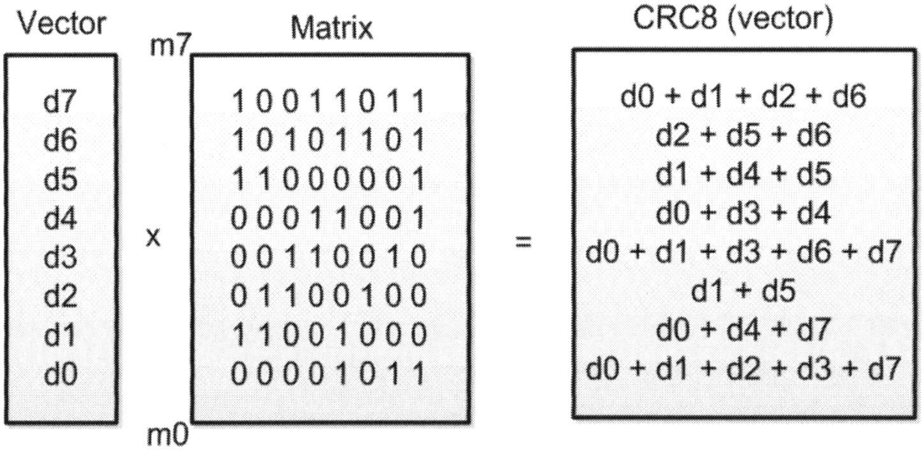

Figure 4. Multiplication by modulo 2 of vector by matrix.

Thus, figure 5 represents the combinational circuit of CRC8 computation by matrix algorithm. There are no multiplexors and constants in the circuit. CRC computation is produced without storage elements using only logic «exclusive OR», which is a feature of matrix algorithm hardware implementation.

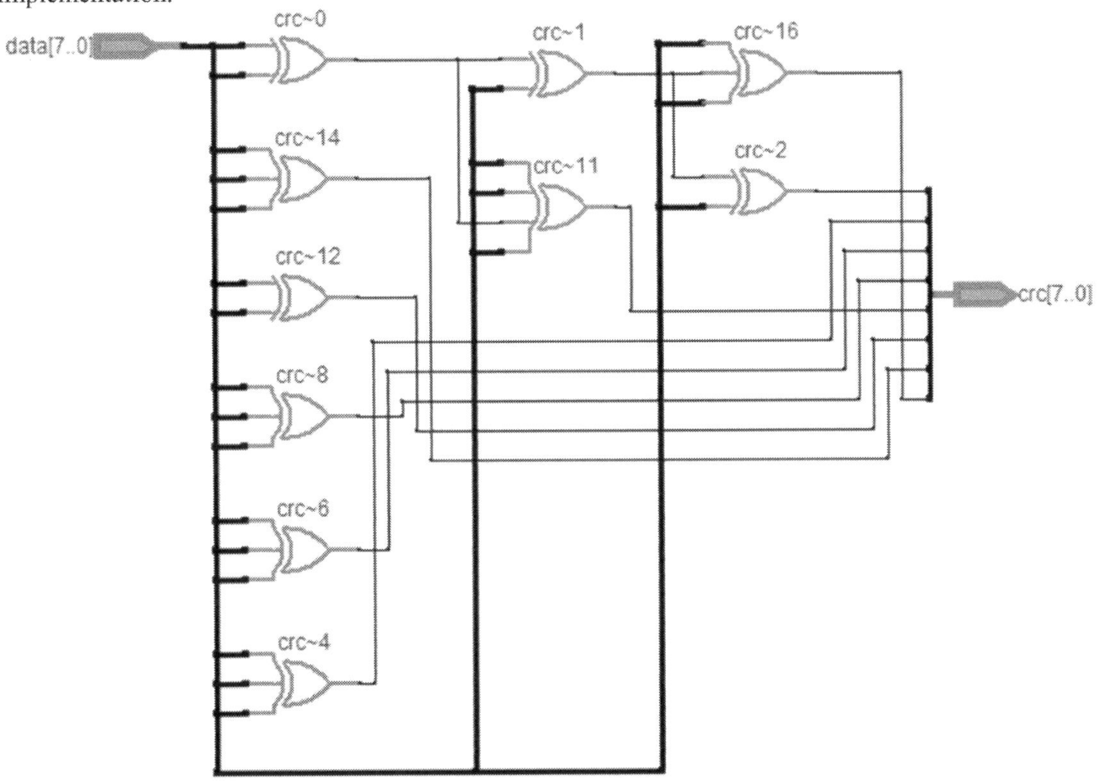

Figure 5. Hardware implementation of CRC8 matrix algorithm using «exclusive OR».

The combinational circuit for CRC32 matrix algorithm is built similarly, but the length of matrix values is increased to 32, which increases the number of logic elements «exclusive OR».

Developed combinational circuits are represented in functional blocks (figures 1, 3) and allow compute CRC for one data byte. CRC computation for data packet is produced by serial communication of functional blocks. Figure 6 shows the diagram of CRC8 matrix hardware implementation such as part of emergency protection system. Each next byte is summed by modulo 2 with computed CRC and result is supplied to next CRC8 block input.

Thus, asynchronous hardware implementation of matrix algorithm has an advantage over table-driven which is reflected in the simplicity of circuit, the absence in the circuit multiplexers and storage elements.

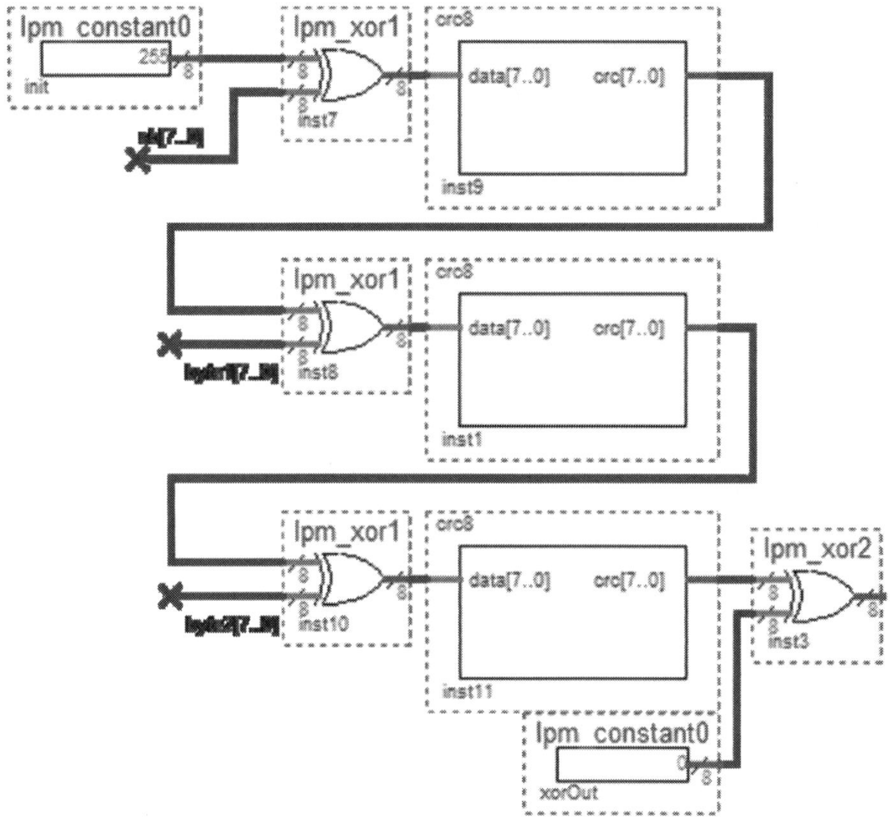

Figure 6. Functional diagram of CRC8 computation for 3 data bytes.

3 Testing of the hardware implementations

Simulation of CRC computation for FPGA Cyclone for different input data was conducted to ensure efficiency of developed devices. Table 1 presents parametric model for CRC8 computation [4]. Figure 7 shows results of CRC8 computation for table-driven and matrix-driven algorithm.

Table 1. Parametric model of CRC8.

Parameter	Value
Name	"CRC 8"
Width	8
Poly	0x9B
Init	0x00
RefIn	False
RefOut	False
XorOut	0x00
Check	0x25

In hardware implementation of CRC algorithms such as in software implementation the parametric model can be changed by changing *Poly, Init, RefIn, RefInOut, XorOut* parameters. However unlike a software implementation in hardware each parameter requires I/O FPGA pins which prevent parametric model control. Usually for a hardware implementation determined parametric model is selected for the device. In the developed devices only the *Init* parameter can be changed.

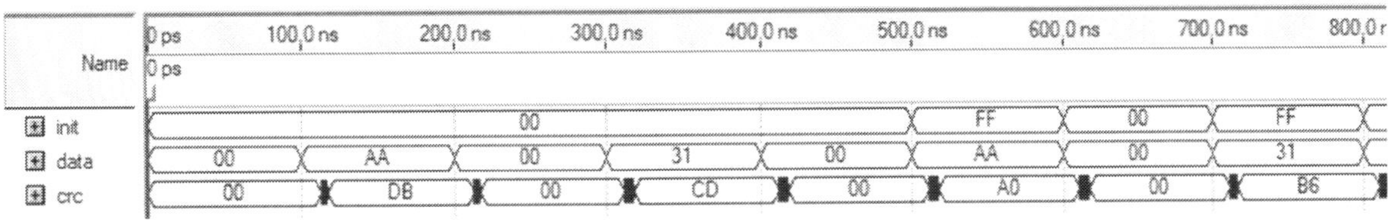

Figure 7. The results of CRC8 device testing.

Thus, values 0xAA and 0x31 are supplied to input *data* for testing device; values 0x00 and 0xFF are supplied to *init* input. Other values of parametric model do not change. Figure 8 shows the result of CRC8 computation by software for the same parametric model.

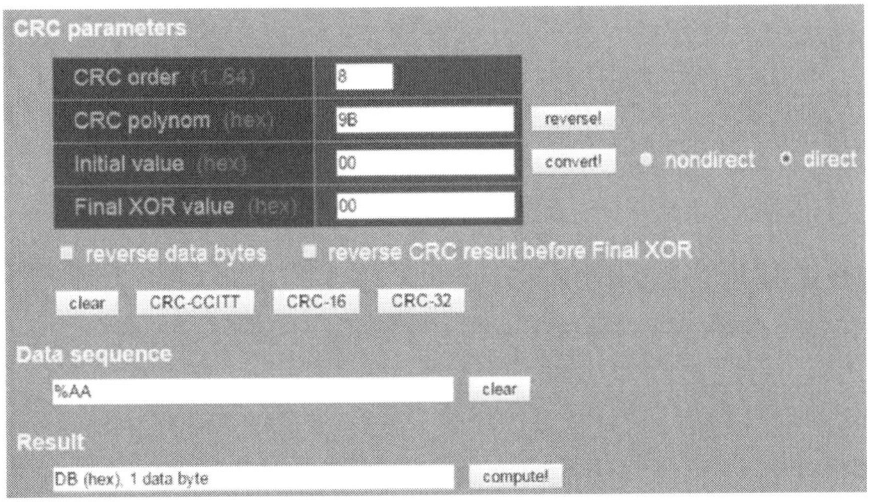

Figure 8. The result of CRC8 computation by software with the same parametric model.

Table 2 presents parametric model for CRC32 computation [4]. Figure 9 shows the results of CRC32 computation for table-driven and matrix-driven algorithms using the parametric model. Figure 10 shows the results of CRC32 computation by software with the same parametric model.

Table 2. Parametric model of CRC32.

Parameter	Value
Name	"CRC 32"
Width	32
Poly	0x04C11DB7
Init	0x00000000
RefIn	False
RefOut	False
XorOut	0x00000000
Check	0x89A1897F

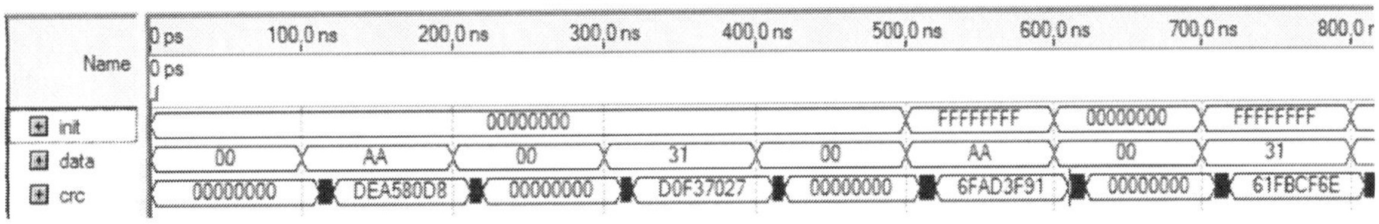

Figure 9. The results of CRC32 device testing.

Values 0xAA and 0x31 are supplied to input *data* for testing device; values 0x00000000 and 0xFFFFFFFF are supplied to *init* input. Other values of parametric model do not change.

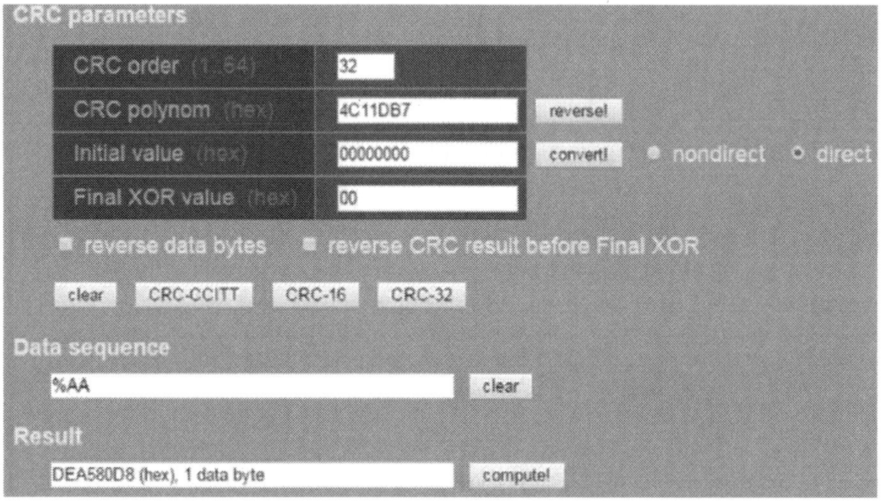

Figure 10. The result of CRC32 computation by software with the same parametric model.

4 Conclusion

The paper discussed hardware implementations of CRC computation algorithms, such as a table-driven which is widely used in data transmission protocols, also a matrix-driven, which allows significantly simplify combinational circuit of CRC computing device. CRC computing device based on FPGA Cyclone which can be used for the satellites configuration memory control and data transmissions modules was developed and tested following parametric model described by R. Williams [4]. Combinational circuit of hardware implementation of matrix-driven algorithm is simpler than table-driven and does not require any complex combinational devices such as multiplexor and storage elements for constant storing. CRC computation circuit in this case is built using only logic «exclusive OR» which gives advantage in hardware implementation. This fact allows embedding CRC computation device based on FPGA in satellites data transmission and configuration modules saving hardware resources for other important modules.

Acknowledgement

The work was performed under the Federal Program «Research and development in priority areas of science and technology Complex of Russia in 2014 - 2020 years» State contract № 14.578.21.0032.

References

1. M. Evgeniy, M. Andrey, *Proceedings of the 9th International Forum on Strategic Technology* (IFOST), 5 (2014)
2. H. Payal, K. Mankar, Intern. J. of Innov. Res. in Adv. Eng. (IJIRAE) **2**, 254 (2015)
3. S. Cherian, N. George, S. Enosh, S. Pillai, Intern. J. of Scien. and Res. Pub. (IJSRP) **4**, 1 (2014)
4. W. Ross, *A Painless Guide to CRC Error Detection Algorithms* (1993)
5. E. Mytsko, *Modern Techniques and Technologies: Proceedings of the 18th International Scientific and Practical Conference of Students, Post-graduates and Young Scientists*, 124 (2012)
6. A. Malchukov, A. Osokin, Y. Bourkatovskaya, *7th Korea/Russia International Symposium on Science and Technology* (KORUS), 189 (2003)

MATEC Web of Conferences 48, 04002 (2016)
DOI: 10.1051/matecconf/20164804002

A Single Information Space for aerospace enterprises based on a Service Oriented Architecture

Ekaterina Vasilyeva[1,a] and Nikolay Markov[1]

[1] *National Research Tomsk Polytechnic University, 634050 Tomsk, Russia*

Abstract. Nowadays aerospace enterprises have a variety of Information Systems (IS) and Software Complexes (SC). Their strategic aim is to integrate all IS and SC, thus to create the Single Information Space (SIS). An approach to create the SIS based on a Service Oriented Architecture (SOA) concept at aerospace enterprises is suggested. It was shown that different tools to implement this approach are available. The SOA model of the SIS of the typical enterprise, which has a custom set of services, IS and SC, was developed. Two modifications of this SOA model are proposed. The first modification is suitable for creation of the SIS on enterprises which have Business Intelligence (BI) systems based on Data Warehouse (DW). Another modification of the SOA-based SIS is important for those enterprises which require to process at least the part of technological and manufacturing data in a real-time mode.

1 Introduction

Nowadays manufacturing enterprises of aeronautical and space branches (hereinafter called as aerospace) are characterized by a huge amount and diversity of issues to solve. Thus, there are many executable Business Processes (BP) which need to be automated. The situation is complicated by several management levels (from a manufacturing sector, a workshop to a management company) which interact in such enterprises. Hence the number of new IS which are implemented to automate routine actions of workers and to support a decision-making on each management level is constantly increasing. Also enterprises use Automated Process Control Systems (APCS) to manage technological processes. Many enterprises have various Laboratory Information Management Systems (LIMS).

Some of Russian aerospace enterprises use or implement Corporate Information Systems (CIS) which are able to automate a work of some enterprise's services and BP. Usually, CIS is presented by a system that controls all enterprise's resources – Enterprise Resource Planning (ERP). However, the experience [1] has shown that ERP system does not allow automating technological processes and some manufacturing processes of the industrial enterprise, so enterprises use a variety of APCS and a system for the manufacturer executive managing which known as Manufacturing Execution System (MES) [2] in addition to ERP system. Some enterprises have not the developed system of ERP class but have an implemented multifunctional MES to control many of enterprise's services, and this MES can act as CIS. Some characteristics of aerospace enterprises (high-tech manufacturing that can be changed in a short time period to release new products; production of unique products or products in small quantities, etc.) determine the necessity to keep on use specialized IS and SC in addition to

[a] Corresponding author : katya-vas@mail.ru

mentioned systems. In most cases such IS and SC are presented by Product Lifecycle Management (PLM) systems, Supply Chain Management (SCM) systems and etc. The large equipment stock, various test stands, unique systems for enterprise's technological tooling at such enterprises leads to necessity of implementation IS to control the technical maintenance and repair of these items of equipment – Maintenance, Repair, and Operations (MRO) system.

The strategic aim of the enterprise which has a large amount of various IS and SC is to create a SIS [3]. The SIS is a set of embedded in the enterprise IS and SC which are interrelated and work according to a common base of principles and rules, providing a convenient interaction between all enterprise's services and employees with a full satisfaction of their information requests. The SIS allows avoiding duplication of input data, to create and use common for whole enterprise classifiers, dictionaries and standards and other similar facilities.

Various methods of integration of IS and SC can be used to create the SIS: data integration, functional services integration and etc. [3]. Nowadays the problem of the creation of the SIS in most cases is solved by data integration with ability to exchange the data between two or more specific IS by using partial integration solutions ("point-to-point" integration). This approach has a number of drawbacks associated with the lack of flexibility and scalability of these solutions. A more general approach to solve the problem at the enterprise level is to use hubs or integration platforms [2, 3]. The most famous modern multifunctional integration platforms are IBM Web Sphere and Microsoft Biz Talk [4]. Along with the versatility and flexibility of these platforms, they are ineffective because of their architecture complexity and the absence of the orientation on specific industries. That's why in recent years industry-based platforms have been developed, for example, for enterprises producing oil and gas [2, 5]. Unfortunately, there are not such specialized platforms in aerospace enterprises.

There is an idea of integration of IS with different architectures and functionality by using a SOA concept in [6], which was succeed. This article deals with the development of this idea, suggested to create the SIS based on SOA concept at aerospace enterprises. Moreover, the SIS may include both existing IS and SC and newly developed and implemented at the enterprise IS and SC, which are presented by a set of services (web services). Tools for creation of the SIS are considered and the SIS SOA model is proposed.

2 Features of creation of the SOA-based SIS for enterprises

We consider the approach to the creation of the SOA-based SIS for manufacturing enterprises, which takes into account the specific requirements for the SIS in the aerospace industry. The basic requirements are:

- to reduce the cost of integration of existing enterprise's IS and SC;
- to reduce the time and cost of development and integration of new IS and software modules into the SIS;
- to increase the speed of applying changes, the list of components of the SIS and the flexibility of BP in the case of rapidly changing business requirements in the aerospace industry (for example, a change in the product mix or the implementation of new technologies).

We show how these requirements can be met through the creation of SIS based on a concept of SOA. SOA is a modular approach to the development of software and IS that based on the use of distributed, loosely coupled and replaceable software components (services), which are equipped with standardized interfaces for communication by standardized protocols [6]. In this case the service is standalone software component (module) with the possibility of self-description which performs a specific business task (BP, sub process, function) and provides some of the functionality by the request to IS or other services. In most cases web services are developed. Loosely coupled services provide a simple and fast adaptation of the SIS to changes in enterprise's BP. The services can work standalone or to be combined with other services to create a new composite applications (IS or SC). The services allow implementing a particular BP by their interaction on the network in a certain sequence. The service should be used according to the convention of the interface. The interfaces are used for providing the service's methods to the outside world and to organize an interaction of services. The service interface contains the description of input parameters and of the result of its work.

Another key component of SOA is a service registry. It is presented by a catalogue and includes descriptions of services (their physical location, versions and validity period; documentation for services). Due to using the service registry SOA has loosely coupled services. The third component of SOA is an Enterprise Service Bus (ESB) [7]. The ESB is connective software for all existing (legacy) IS and SC on the enterprise and all new developed services to provide centralized and unified event-driven messaging between IS, SC and new enterprise's services. All settings for the processing and sending of messages are usually concentrated in a single point of ESB. It means that in case of change of any IS, SC or service connected to the bus there is no need to reconfigure other systems and services of the SIS. The ESB is responsible for a number of classical tasks of integration of IS, SC and services:

- message routing;
- conversion of message formats and protocols;
- transaction support;
- providing security, auditing, logging.

Finally, another component of the SOA is Business Process Management (BPM) system. BPM system is software providing opportunities to start, coordinate, manage, administrate and debug the BP. The BP is deemed to be a sequence of logically related actions that transform input data into results or output data. In other words, BP determines a startup sequence for IS, SC and services to solve specific business task. Management of the startup sequence is called orchestration. BPM system also controls the choreography in the case of the exchange of messages between IS, SC and services.

Summarizing the brief analysis of the components of SOA and its main features, we can assume that SOA is a promising model of interaction between the SIS components (legacy IS, SC and new services). This model allows the components to communicate with each other through specified interfaces. The interfaces are not dependent on hardware platforms, operating systems and programming languages that were used to develop components of the SIS on the enterprise. This allows components to communicate with each other by the same standard and versatile way. This aspect of using interfaces that are independent of the environment and the platform is called "loose coupling" model. This model provides greater flexibility and adaptability of the SIS in the changing requirements of the enterprise as the replacement or upgrading a component of the SIS will not affect the others. Therefore, the SOA-based SIS will meet the basic requirements for the SIS in the aerospace.

3 Tools for implementation the SOA-based SIS

We made a brief overview of tools and technologies to create the SOA-based SIS on the enterprise. Following two most common technologies are used to describe the functions of services and messages [7]:

- Web Services Description Language (WSDL) is a language for web-services and access to them; the technology of development of such services divides the description of services into the description of the interface and shell: the interface describes what must contain the request and the shell determines the data and transport protocols;
- Simple Object Access Protocol (SOAP) is a protocol for exchange structured messages that is used for description of format of received and sent messages.

The ESB should be presented by one of service buses of known software companies such as Oracle, IBM, Microsoft and etc. More full ESB functionality is presented in Oracle ESB, Open ESB and IBM Web Sphere ESB [7]. Most ESB comes with a large number of pre-configured adapters: adapters for ERP systems (SAP R/3, Oracle eBusiness Suite, 1C, MS Ax 2009), adapters for databases and etc.

The BPM-system describes the coordination of BP using the Business Process Execution Language (BPEL). The description of BP on BPEL is compiled into an executable code of the BPM system that allows to recover described BP from available IS, SC and services [8]. In recent years a more advanced language for modeling BP was invented – the language BPMN (Business Process Model and Notation) [9]. BPMN specification allows to present BP as diagrams and to convert them to BPEL. There is an example of such diagram on the figure 1 which describes the BP "Inspection of workshop's production equipment". Sub processes "Forming an inspection list", "Filling of electronic diagnostics journals" and "Filling of electronic operations journals" are executed in the MRO system, sub process "Informing of related units" can be implemented in Electronic Document Management System (EDMS), sub process "Responding to a failure" can be implemented by MES with creation of the task to do unscheduled work and making orders of required material resources in the ERP system. The BPM system which is based on the diagram is aimed to coordinate all of the mentioned IS. The BPM system can be presented by widely known systems: Orchestra (supports both languages), Bonita BPM (language supports BPMN), etc.

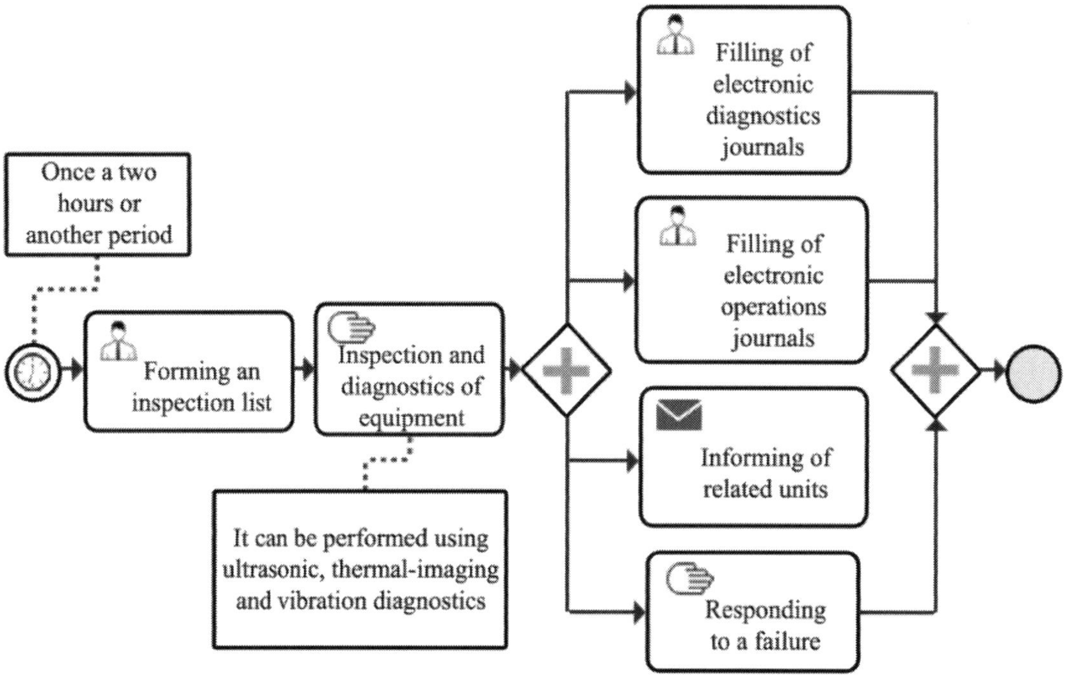

Figure 1. BPMN diagram of the BP "Inspection of workshop's production equipment".

4 SOA models of the SIS of enterprises

The approach to the creation of the SOA-based SIS on aerospace enterprises is described above. We reviewed features of it in context of the typical aerospace enterprise which has standard set of IS, SC and services. The SOA model is based on ESB (figure 2) which usually presented by one of described buses. The most of legacy IS and SC (PLM system, ERP system, SCM system, MES, EDMS, MRO system and etc.) have adapters to communicate by the ESB. Some adapters are standard and they are supplied with the ESB, while others should be developed according to the specification of the concrete IS. Special OPC adapters (based on the open standard OLE for Process Control) are implemented into all enterprise's APCS and LIMS to provide an access to technological data and results of laboratory research through some IS. Widely used specifications OPC DA (Data Access) and OPC XML DA allow real-time exchanging data with programmable logic controllers, numeric control and etc. Typical enterprises also must have a system to control referenced data (SCRD). Using the SOA model has some benefits, for example, BPM system is able to perform EDMS functions. Services (from 1 to N) for new IS and SC are created accordingly their interface specifications.

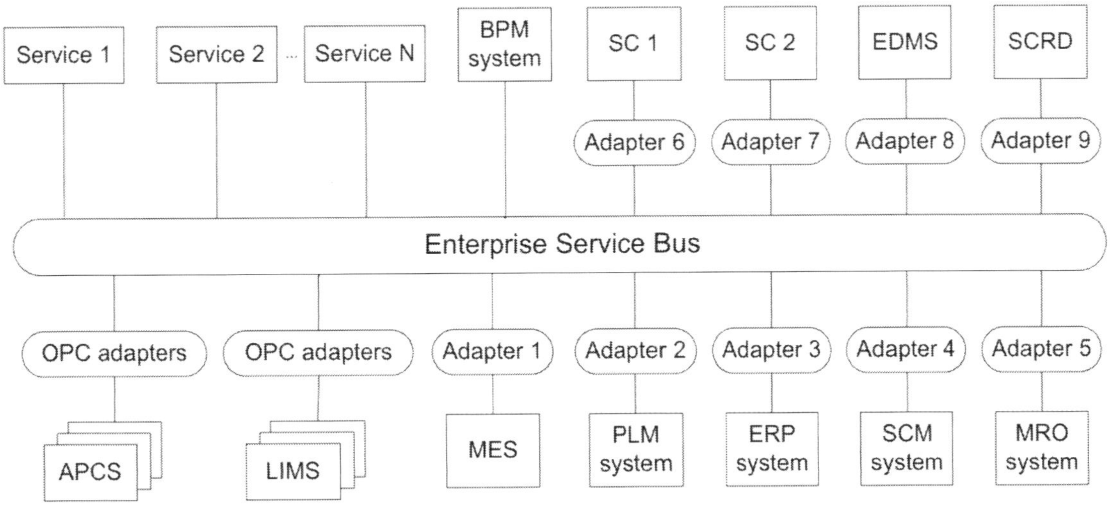

Figure 2. The SIS architecture of the typical enterprise.

The SIS of untypical aerospace enterprise additionally can contain BI systems based on DW. Moreover, in case of BI analysis of big data from APCS, LIMS and MES the SOA model of the SIS of the typical enterprise will be complicated by additional system bus for BI systems and DW. The additional bus provides the data and messages exchange between BI systems and DW that allows decreasing the usage of the ESB.

The SOA model of the SIS of the typical enterprise should be modified by additional Manufacturing Services Bus (MSB) to provide an opportunity of real-time technological data processing (according to the concept "Manufacture 2.0" [7]). In this case SOA model from the figure 2 should be integrated with manufacturing SOA model and some IS and SC will be moved to the MSB. It allows organizing the technological and manufacturing data and messages exchange in a real-time mode by facilities of the manufacturing SOA model.

5 Conclusion

Nowadays aerospace manufacturing enterprises are characterized by a huge amount of IS and SC to automate a variety of BP. At the same time there is a necessity to create the SIS that will provide a convenient interaction between all enterprise's services and employees when they interact with IS and SC. The SIS is a set of embedded in the enterprise IS and SC which are interrelated and work according to a common base of principles and rules. The SIS allows avoiding duplication of input data, to create and use common for whole enterprise classifiers, dictionaries and standards and other similar facilities.

The approach of creating the SIS based on SOA, which is able to meet requirements for the SIS of aerospace enterprises, is presented in the article. One of such requirements is a flexibility of BP in the case of rapidly changing business requirements in the aerospace industry, and the main requirement for the automation process of BP is a high speed of applying changes to the SIS components. It is shown that creation of the SOA-based SIS will provide the required for aerospace enterprises flexibility in the case of implementation of new services or removing legacy IS and SC. This can be achieved through the organization of services, IS and SC interaction through the ESB. An analysis of tools and technologies for development of SOA models of the SIS has shown that all necessary tools for this are available. The ESB can be presented by one of service buses of known software companies. It is suggested to use one of widely known BPM systems which support BPEL and (or) BPMN.

The SOA model of the SIS of the typical enterprise, which has a custom set of services, IS and SC, was developed. This SOA model can be modified according to changing requirements of the enterprise. The modification for enterprises which have BI systems based on DW is presented. This modification has an additional bus for BI systems and DW which is integrated with the ESB. The additional bus provides the data and messages exchange between BI systems and DW that allows decreasing the usage of the ESB. Another modification of the SOA-based SIS is important for those enterprises which require to process at least the part of technological and manufacturing data in a real-time mode. In this case the SOA model of the SIS of the typical enterprise integrates with the manufacturing SOA model which has the MSB. Using the manufacturing SOA model supposes moving IS, which are working in a real-time mode (APCS, LIMS), to the MSB. It allows organizing the technological and manufacturing data and messages exchange in a real-time mode by facilities of the manufacturing SOA model.

References

1. J. Kletti (Ed.) *Manufacturing Execution Systems – MES*, (Berlin Heidelberg: Springer-Verlag, 2007)
2. A. V. Kudinov, N. G. Markov, *Problemy avtomatizacii proizvodstva gazodobyvajushhih kompanij* (Tomsk: izdatelstvo Tomskogo politehnicheskogo universiteta, 2012)
3. N. G. Markov, Informacionnoe obshhestvo **3**, 53 (2014)
4. Microsoft BIZ Talk Server. URL: https://www.microsoft.com/en-us/server-cloud/products/biztalk/ (2016)
5. V. Veyver, A. Kudinov, N. Markov, International Journal of Computer Applications **5**, 14 (2012)
6. M. Juric, *SOA approach to integration* (Birmingham: Packt Publishing Ltd, 2007)
7. A. Boyd, D. Noller, P. Peters, *WP 27. SOA in Manufacturing Guidebook* (MESA International, IBM Corporation and Capgemini co-branded white paper, 2010)
8. W. H. Utomo, T. Wellem, JATIT **62**, 364 (2014)
9. BPMN – model biznes-processov i notacija. URL: www.elma-bpm.ru (2016)

MATEC Web of Conferences **48**, 04003 (2016)
DOI: 10.1051/matecconf/20164804003

Speckle pattern processing by digital image correlation

Fedor Gubarev[1,2,a], Lin Li[1], Miron Klenovskii[1,3] and Anatoliy Glotov[1]

[1] *National Research Tomsk Polytechnic University, 634050 Tomsk, Russia*

Abstract. Testing the method of speckle pattern processing based on the digital image correlation is carried out in the current work. Three the most widely used formulas of the correlation coefficient are tested. To determine the accuracy of the speckle pattern processing, test speckle patterns with known displacement are used. The optimal size of a speckle pattern template used for determination of correlation and corresponding the speckle pattern displacement is also considered in the work.

1 Introduction

The progress of space exploration and space technologies is inseparably linked with the development of new methods and means of remote obtaining information about the objects and processes. The objects under control can be living organisms as well as various machine parts during the process of their manufacturing or their operation. Promising technics of remote control and at the same time nondestructive testing are the optical methods, and specifically methods based on the use of coherent radiation [1-6].

One of the most promising technics for the measurement of microdisplacements and deformations is a method of digital speckle correlation [6-8]. This method, which has found wide application in nondestructive testing of materials [9-11], is also a promising method of diagnostics in biology and medicine [7, 12-15]. The method of digital speckle correlation is also suitable for the remote measurement of microdisplacements and speed of various processes [15-17]. In [15] the authors realized audio signal extraction from vibrating objects located at a distance up to 100 m by means of the digital speckle correlation method. This fact points to the great prospects of the technique. For instance, in a case of radio communication failure this feature can allow to realize a possibility to hear an astronaut when he/she is outside a space vehicle just directing a laser beam on the spacesuit.

The essence of the digital speckle correlation technique consists in illuminating the object under investigation by coherent radiation with simultaneous recording speckle patterns formed by light scattered from the object and subsequent their processing by means of the method of digital image correlation. In [15] is shown that when optical system being used is properly configured, the displacement of an object under observation results in proportional transverse shift of the formed speckle pattern. This fact allows to detect the displacement as well as vibration of the object. The displacement of objects under control can be caused by various factors which can be both desirable and undesirable. Having established a correlation between the change in the formed speckle pattern

[a] Corresponding author : gubarevfa@tpu.ru
[2] *V.E. Zuev Institute of Atmospheric Optics, Russian Academy of Sciences, 634021 Tomsk, Russia*
[3] *Institute of Electron Physics, National Academy of Sciences of Ukraine, 88017 Uzhgorod, Ukraine*

and the change in a controlled process, it is possible to realize a complete control of the process by implementing the appropriate control action (a feedback).

One of the main requirements for the method is the high reliability and accuracy of determining the displacement of an object of interest. The aim of the current research is to study the method of speckle pattern processing by digital image correlation. The optimal size of a speckle pattern template used for determination of correlation and corresponding the object displacement is also investigated in the work.

2 Description of speckle pattern processing

When a rough surface of an object is illuminated by a coherent radiation, the beams reflected from the surface interfere and as a result spatially distributed speckle patterns are formed. Any displacement of an object under observation causes corresponding change in the spatial distribution of obtained speckle patterns. By establishing these changes one can determine the displacement of the object surface. Moreover, determining the time-resolved displacement of the object surface one can measure the vibration of the surface [18].

The determination of a surface displacement can be easily carried out by analysis of speckle pattern shift by means of digital image correlation. Let us consider the bottom lines of the method. The speckle patterns recorded before and after displacement of the surface are needed to carry out the analysis. The recorded digital speckle images are converted into matrices contained values of pixels intensity.

Figure 1 illustrates the change in position of a speckle after displacement. As one can see initially the position of a speckle has coordinates (x_1, y_1) but after displacement it changes to the (x_2, y_2). Let the intensity in the some part of the speckle pattern before the displacement is $f(x, y)$ while its intensity after the displacement is $g(x, y)$. If in the image obtained after displacement one can find a region g correlated best with the initial speckle f, then one can conclude that speckle is displaced from the position of f to the position of g.

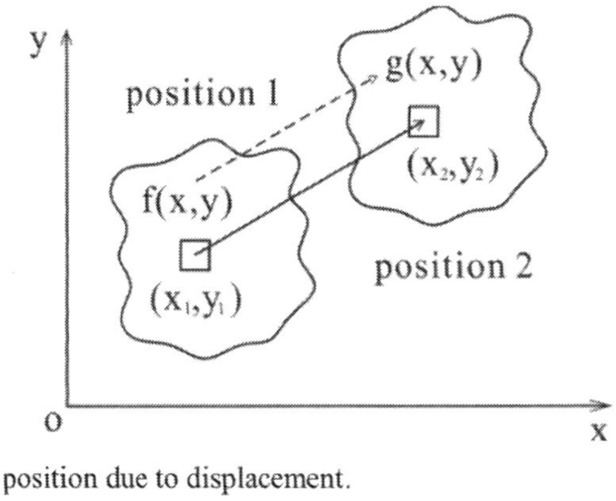

Figure 1. Changing a speckle position due to displacement.

To find the position of g it has been realized an algorithm in the MATLAB software environment. The main idea of the algorithm consists in finding the location of the correlation coefficient maximum. The algorithm contains the following three basic steps.

1. In the first image a part of the speckle pattern (hereinafter referred to as template) is chosen which then is searched in the next images. If the chosen template is of too small size, it is impossible to localize its position in the next images because of lack of unique information in the template of small size. If the chosen template is of too big size, it takes a lot of time for the processing of images. In [8] is mentioned that the optimal size of the template is found over the range 31×31 pixels to 61×61 pixels. However, no detailed explanations regarding this conclusion are given. Therefore, in the current work the special attention is paid to the question about finding the optimal size of the template.

2. The algebraical expression for calculating the correlation coefficients is chosen. The most widely used formulas of the correlation coefficient are the following [9, 10, 19-22]:

$$C_1 = \frac{\sum_{i=1}^{n}\sum_{j=1}^{m}[f(x_i, y_i)\cdot g(x_i, y_i)]}{\sqrt{\sum_{i=1}^{n}\sum_{j=1}^{m} f^2(x_i, y_i)\cdot \sum_{i=1}^{n}\sum_{j=1}^{m} g^2(x_i, y_i)}} \tag{1}$$

$$C_2 = \frac{\sum_{i=1}^{n}\sum_{j=1}^{m}[f(x_i, y_i) - \overline{f}]\cdot[g(x_i, y_i) - \overline{g}]}{\sqrt{\sum_{i=1}^{n}\sum_{j=1}^{m}[f(x_i, y_i) - \overline{f}]^2 \cdot \sum_{i=1}^{n}\sum_{j=1}^{m}[g(x_i, y_i) - \overline{g}]^2}} \tag{2}$$

$$C_3 = (C_2)^2 = \frac{\left\{\sum_{i=1}^{n}\sum_{j=1}^{m}[f(x_i, y_i) - \overline{f}]\cdot[g(x_i, y_i) - \overline{g}]\right\}^2}{\sum_{i=1}^{n}\sum_{j=1}^{m}[f(x_i, y_i) - \overline{f}]^2 \cdot \sum_{i=1}^{n}\sum_{j=1}^{m}[g(x_i, y_i) - \overline{g}]^2} \tag{3}$$

Where, $\overline{f}$ is the average intensity of the reference (initial) speckle pattern (template); $f(x_i, y_i)$ is intensity of its pixel with coordinates (x_i, y_i); $\overline{g}$ is the average intensity of the speckle pattern obtained after displacement; $g(x_i, y_i)$ is intensity of its pixel with coordinates (x_i, y_i).

The main requirements for the formula of the correlation coefficient calculation are reliability, noise immunity, and speed of calculating. In practice, the formula (2) is the most widely used. It should be mentioned that in this step the location of the correlation coefficient maximum is found with accuracy within +/– 0.5 pixels.

3. The correlation coefficients around the found maximum value are chosen to carry out their interpolation with a step of 0.01 pixels or even less. Thus, a more accurate location of the correlation coefficient maximum is found.

3 Testing the method of speckle pattern correlation

In the current work testing of the realized algorithm is carried out by means of the test speckle patterns with known displacement. In [23] a method for producing high-precision experimentally subpixel shifted images has been created. The author [23] produced a set of speckle patterns shifted with a step of 0.1 pixels. There are available two sets of speckle patterns with high and low contrast [24] which are used in the current work for determining the accuracy of the algorithm. Figure 2 exhibits the trajectory of speckle pattern shifting and its determination by the algorithm under testing.

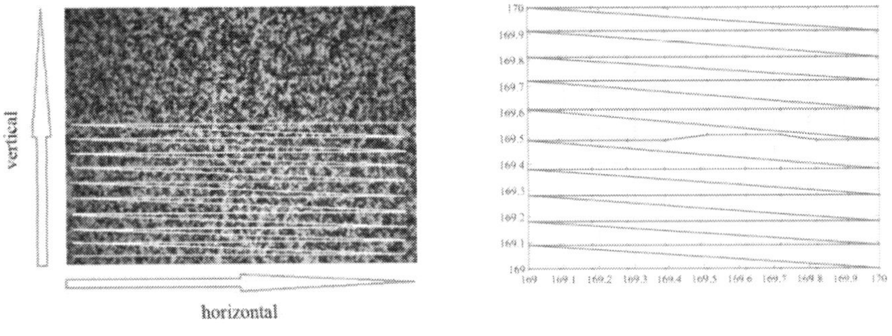

a *b*

Figure 2. A speckle pattern and the trajectory (yellow curve) of its shifting (*a*) and the trajectory of the speckle pattern displacement determined by the algorithm under testing (*b*).

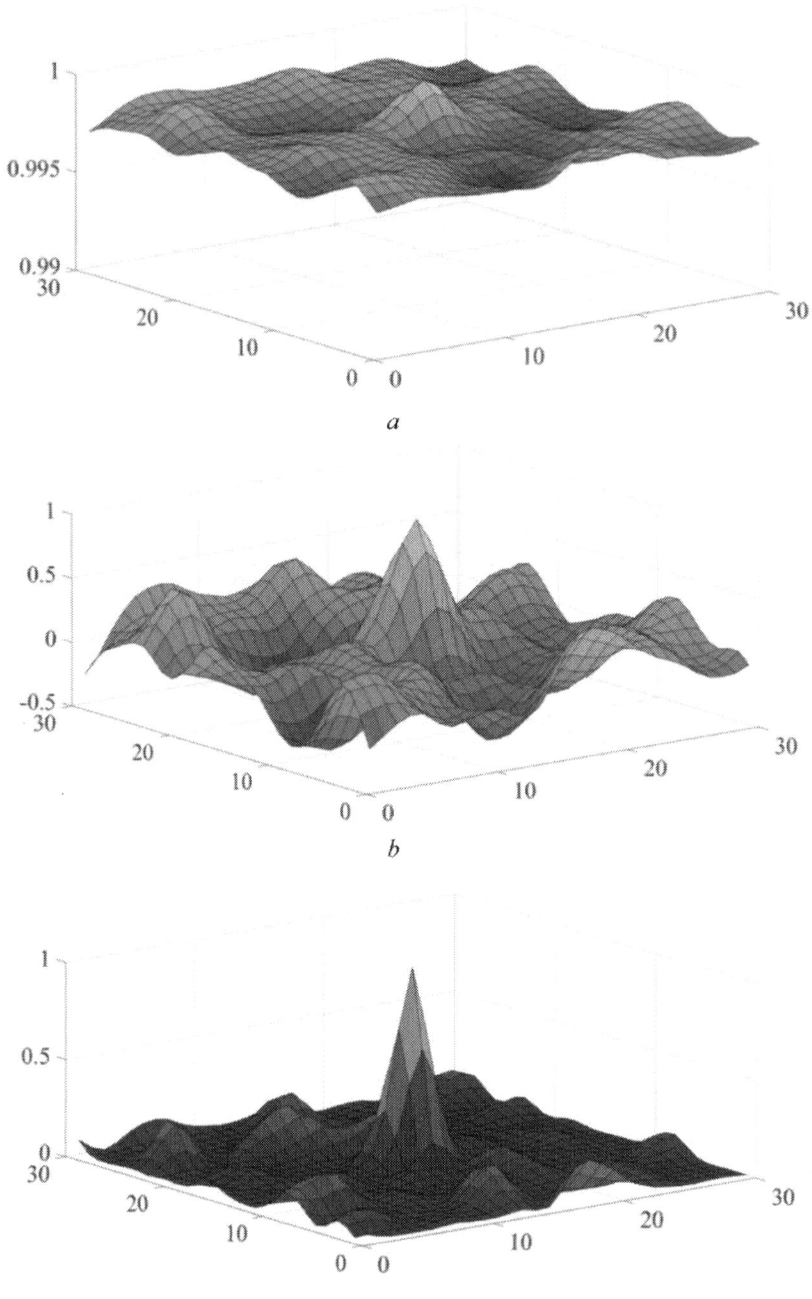

Figure 3. Distribution of the correlation coefficient calculated by the formulas: a) – (1); b) – (2); c) – (3).

Figure 3 presents distribution of the correlation coefficient calculated by the formulas C_1–C_3 for the high contrast speckle pattern borrowed from [24]. One can notice that the distribution of the correlation coefficient calculated by the formula (1) is not characterized by unimodality and clear peak. As a result the formula (1) cannot ensure the reliable localization of the correlation coefficient maximum. The calculation carried out by the formula (2) gives a lot of peaks but the main peak is already stood out well against the others. The calculation performed by the formula (3) results in the best unimodality. The difference between the main peak and the others is noticeably more. However, the question about the formula ensured the best final result is not obvious. To clear up the question the test calculations were carried out.

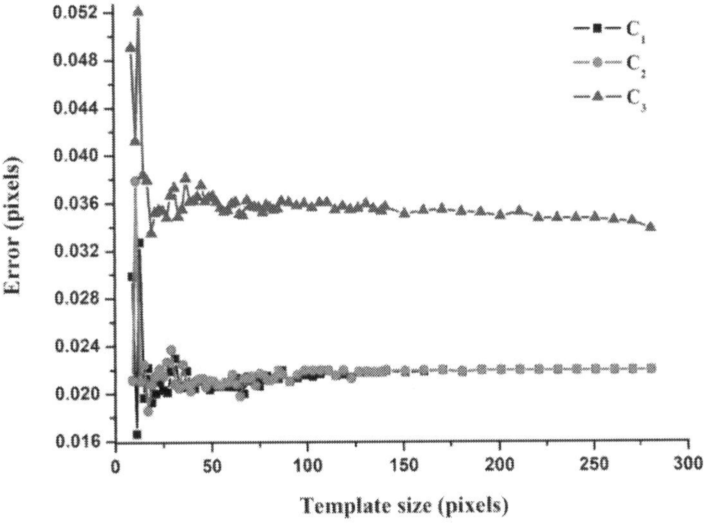

a

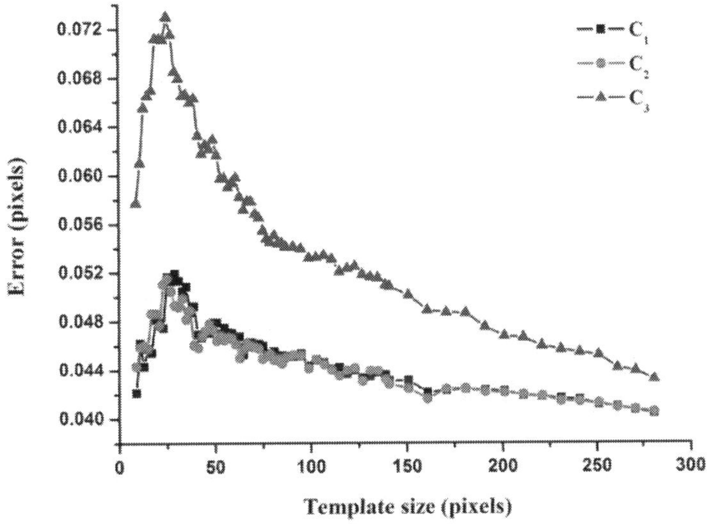

b

Figure 4. Dependence of the average absolute error of displacement determination on the template size (side length) for the different formulas of correlation coefficient.

Figure 4 shows dependence of the average absolute error of displacement determination on the template size for the different formulas of correlation coefficient. The calculation by the formulas (1) and (2) gives almost the same rusults (figure 4). The average absolute error of displacement determination by the formulas (1) and (2) amounted to 0.022 pixels for the high contrast speckle pattern set and 0.044 pixels for the low contrast speckle pattern set. As one can see the calculation by the formula (3) gives almost twice increase in the error of displacement determination dispite the presence of clear main peak (figure 3c). One can suppose that this is due to increasing the difference in the correlation coefficients around the main peak which leads to deterioration of the interpolation quality and as a result to the accuracy degradation of the correlation maximum localization.

It should be also noted that when the template size is less than 30×30 pixels the accuracy of displacement determination is not stable. To get the reliable results the template size should be more than 30×30 pixels. In the case of low contrast speckle patterns the bigger template size gives the better accuracy (figure 4b).

4 Conclusion

Studying the method of speckle pattern processing based on the digital image correlation was carried out. The three formulas of correlation coefficients were investigated. The first two formulas were turned out to give the best accuracy of determining the speckle pattern displacement. The formulas ensured the average absolute error of displacement determination about 0.022 pixels for the high contrast speckle pattern set and 0.044 pixels for the low contrast speckle pattern set. However, there is a suspicion concerning the reliability of the first formula. One can suppose that in a case of speckle patterns with a noise the first formula can give unreliable results because of low noise immunity.

The optimal size of a speckle pattern template needed for determination of correlation and corresponding the speckle pattern displacement was also investigated in the work. When the template size is less than 30×30 pixels the accuracy of displacement determination was turned out to be not stable. To get the reliable results the template size should be more than 30×30 pixels.

References

1. I.P. Sukharev, I.P. Sukharev, *Experimental Methods for the Study of Deformation and Strength* (Mechanical Engineering, Moscow, 1987)
2. G.S. Evtushenko, D.V. Shiyanov, F.A. Gubarev, *Metal Vapor Lasers with High Pulse Repetition Rates* (Tomsk Polytechnic University Publishing House, Tomsk, 2010)
3. F.A. Gubarev, M.S. Klenovskii, Lin Li, IOP Conf. Series: Materials Science and Engineering **81**, 012116 (2015)
4. F.A. Gubarev, M.V. Trigub, M.S. Klenovskii, Lin Li, G.S. Evtushenko, Appl. Phys. B. **122**, 1 (2016)
5. M. Locatelli, E. Pugliese, M. Paturzo, V. Bianco, A. Finizio, A. Pelagotti, P. Poggi, L. Miccio, R. Meucci, P. Ferraro, Opt. Express **21**, 5 (2013)
6. J.W. Goodman, *Speckle Phenomena in Optics: Theory and Applications* (Roberts and Company Publishers, Englewood, 2007)
7. V.V. Tuchin, *Lasers and Fiber Optics in Biomedical Research. 2 edition.* (Fizmatlit, Moscow, 2010)
8. G. Jin, L. Meng, J. Chen, S. Ma, J. Zhang, Exp. Mech. **21**, 689 (2006) (in Chinese)
9. D. Zhang, X. Zhang, G. Cheng, Exp. Mech. **39**, 62 (1999)
10. H. Lu, P.D. Cary, Exp. Mech. **40**, 393 (2000)
11. J. Gonzalez, W.G. Knauss, J. Mech. Phys. Solids **46**, 1981 (1998)
12. V.V. Tuchin, J. Biomed. Opt. **4**, 106 (1999)
13. M.A. Vilensky, D.N. Agafonov, D.A. Zimnyakov, V.V. Tuchin, R.A. Zdrazhevskii, Quantum Electron **41**, 324 (2011)
14. Y. Beiderman, N. Burshtein, M. Teicher, V. Mico, Z. Zalevsky, J. Biomed. Opt. **15**, 061707 (2010)
15. Z. Zalevsky, Y. Beiderman, I. Margalit, S. Gingold, M. Teicher, V. Mico, J. Garcia, Opt. Express **17**, 21566 (2009)
16. B.R. Durig, W.H. Peters, M.A. Hammer, Proc. SPIE **0954**, 438 (1989)
17. N.A. Russo, J.A. Pomarico, E.E. Sicre, D. Patrignani, L. Pasquale, Opt. Commun. **169**, 23 (1999)
18. Lin Li, F.A. Gubarev, M.S. Klenovskii, A.I. Bloshkina, International Siberian Conference on Control and Communications (SIBCON) (to be published)
19. K. Okamoto, Y.A. Hassan, W.D. Schmidi, Exp. Fluids **19**, 342 (1995)
20. T. Roesgen, R. Totaro, Exp. Fluids **19**, 188 (1995)
21. L. Gui, W. Merzkirch, Exp. Fluids **28**, 36 (2000)
22. W.S. Schreier, J.R. Braasch, M.A. Sutton, Opt. Eng. **39**, 2915 (2000)
23. P.L. Reu, Exp. Mech. **51**, 443 (2011)
24. Digital image correlation challenge. URL: http://www.sem.org/dic-challenge (2016)

MATEC Web of Conferences 48, 05001 (2016)
DOI: 10.1051/matecconf/20164805001

Computational approach to assess postural tests under microgravity conditions

Konstantin Brazovskii[1,a], Vasilii Fokin[2], Ivan Tolmachev[2], Jacov Pekker[2] and David Hachaturyan[1]

[1] National Research Tomsk Polytechnic University, 634050 Tomsk, Russia
[2] Siberian State Medical University, 634050 Tomsk, Russia

Abstract. Human vestibular system is extremely fragile and rapidly degrades when gravity is significantly lower than the gravity of Earth. Unfortunately, current balance assessment tools, including state-of-start computerized centre-of-mass measurement platforms, cannot be used in space because of the low gravity. Thus, a problem of developing new methods to assess postural stability under microgravity conditions is very important. This paper describes a distance-based criterion to estimate multidimensional measurements and its application for postural tests assessment. It has been shown that proposed criterion may be effectively used to estimate the equilibrium function by means of low-end hardware, which does not rely on the gravitational force. The criterion was compared against several known indexes and its ability to distinguish disorders of the balance has been studied.

1 Introduction

Human vestibular system demonstrates substantially disturbed behaviour under microgravity conditions and should be monitored constantly during long term space travels. The influence of microgravity on human's sense of balance has been studying for many years, and eventually it became clear that maintaining the equilibrium function in appropriate functional state is extremely important to prevent development of serious disorientation and neurological disorders.

An ability to balance and maintain postural stability is an essential diagnostic criterion of a wide range of disorders. A great variety of tests, functional and quantitative as well, has been developed since the Romberg manoeuvre was introduced in the middle of the XIX century. The general idea of conventional postural tests data processing methods is to estimate statistical parameters of movements under certain conditions, and then summarize the raw values into scores. This approach has been clinically validated and, that is also important, is based on a quite simple numerical procedure with clear interpretation.

The current "gold standard" quantitative test is an analysis of the centre of pressure (COP) by means of high precision computerized force plates [1]. The COP is quite sensitive to Parkinson's disease (PD) [2], even in an early stage of PD, and also to other balance disorders. Unfortunately, it has been reported [3] that the COP method has some disadvantages, such as a large size of equipment, which should be properly installed in stationary conditions and thoroughly calibrated before use. One more issue of the posturography is a requirement that all force platforms intended for clinical

[a] Corresponding author : bks_2005@ mail.ru

applications should satisfy the standards of the International Society for Posture and Gait Research (ISPGR), it means that the devices must be able to measure movements and displacements with ±0,1 mm accuracy and 0,05 mm resolution. Such a high level of metrological characteristics is achievable, but for a high price. As a result, the computerized posturography (CPG) is very useful but relatively expensive method, which can be used effectively to resolve some specific clinical issues rather than on a daily basis. Unfortunately, widely used conventional hardware to evaluate a functional state of the equilibrium sense is strongly depended on the precise value of the gravitational force, thus cannot be used under microgravity conditions.

Reportedly, traditional neurological tests do not work in space because they are tightly linked to gravity [4]. The experimental setup without dependency on the precise value of gravity, proposed in [4], demonstrates impressive diagnostics capability for the price of more complex design and sophisticated data processing techniques.

Taking into account possible consequences of postural instability and balance disorders during long term space travels, the development of valid balance assessment tools usable in space, is undoubtedly important [4]. Recently quite a few hardware tools and computational methods to collect and process data of postural tests have been introduced; primarily for aerospace applications, which means high price and limited availability. The objective of the presented study, inspired by the current aerospace technologies, is to develop a quantitative method to assess maintaining postural stability during functional tests, which could be used under low gravity conditions.

2 Study design

The study was conducted in two groups: the first, reference group, consisted of 9 healthy subjects; the second group was comprised of 9 patients with minor neurological deficits and negative Romberg's test.

Functional postural tests were used along with a video-based markless motion capture technique. The protocol included Romberg's test that followed by the Balance Evaluation Systems Test (miniBESTtest, [4]). The Cartesian coordinates of the featured points were captured by means of Microsoft Kinect hardware. The "Brekel Kinect free" software was used to preprocess and collect raw data.

Finally, our new numerical distance-based criterion was calculated. After that the criterion was compared against the scores of the BESTtest and known indexes, such as average speed and patented Equilibrium Function Quality (EFQ) [6]. Statistical hypotheses were tested by Mann–Whitney U nonparametric test (MW test), Spearmen rank correlation was used as well.

2.1 Materials and methods

The theoretical details of the criterion were published in [7]. Briefly, the data processing pipeline can be given as follow:

1. First of all, time series vectors $\mathbf{x}(t)$, t=0,. . .,T have to be measured for each subject during the period of time with length T. Here $\mathbf{x}(t)$ is a multidimensional value, for example, the 3D coordinates of a certain point.

2. The whole interval $[0,T]$ then is divided into k fragments with length ΔT. The average values $\mathbf{x}_j$, j=0,. . ., k are calculated inside the $[t,\ t+\Delta T]$ intervals (1):

$$\mathbf{x}_j = \frac{1}{n}\sum_{i=1}^{n}\mathbf{x}(t_i) \quad (1)$$

where n is a number of the measured values inside the interval.

3. The averaged values are normalized (2) and put into the N x k state matrix S, N is a number of subjects in a certain group.

$$\mathbf{x}_{j,norm} = \frac{\mathbf{x}_j - \mathbf{x}_{\min}}{\mathbf{x}_{\max} - \mathbf{x}_{\min}}, \quad j = 1,k$$
$$\mathbf{x}_{\min} = \min_{j=1,k}\{\mathbf{x}_j\}, \quad \mathbf{x}_{\max} = \max_{j=1,k}\{\mathbf{x}_j\} \tag{2}$$

4. All averaged and normalized values of the healthy subjects $\mathbf{x}_v$ are merged into the reference matrix $\mathbf{S}_0$ with size N_0, and then a covariance matrix **C** with k x k size is calculated.

5. Finally, the distance-based dimensionless criterion Id is calculated (3), using the averaged and normalized values $\mathbf{x}_u$ obtained from the patients.

$$Id(\mathbf{x}_u) = \frac{1}{2k}\left[\frac{1}{N_0}\sum_{v=1}^{N_0}(\mathbf{x}_u - \mathbf{x}_v)\mathbf{C}^{-1}(\mathbf{x}_u - \mathbf{x}_v)\right] \tag{3}$$

where $\mathbf{x}_v$ are the reference vectors from the reference group, and $\mathbf{x}_u$ are the test vector obtained from a certain patient, $\mathbf{C}^{-1}$ is an inverse covariance matrix. In other words, our criterion is the averaged squared Mahalanobis distance between the certain vector and all vectors from the reference set.

Also we used EFQ test, which is based on a vector statokinesigram and includes calculation of COP's velocity at each moment of time. After that a cumulative distribution function of the velocity distribution was approximated. The actual value of EFQ is relative area under the cumulative curve, which is given as a percentage.

Usually, the conventional methods are able to evaluate movement of a single point (centre of pressure, for example) only. In contrast, the Id is not limited by the dimension of the measured vectors and almost any reasonable length is appropriate. In order to make the developed criterion comparable to the common ones, a point corresponding to the center of mass was chosen. Vertical displacement (z coordinate) was omitted. The data were recorded during Romberg's test with 15 samples/second rate and estimated spatial resolution about one millimetre. The overall period of time was T=10sec, ΔT=2sec, n=30.

Table 1. Individual parameters for the studied subjects.

Subject	MiniBEST test scores	Average speed *Me* [Q0,25-Q0,75] m/sec	EFQ, %	Id
R1	27	0,103 [0,056-0,159]	81,34	1,63
R2	28	0,015 [0,008-0,021]	73,83	1,37
R3	28	0,057 [0,033-0,096]	85,23	1,21
R4	28	0,013 [0,008-0,021]	81,01	1,15
R5	27	0,034 [0,019-0,063]	83,89	1,30
R6	28	0,021 [0,010-0,032]	76,04	1,35
R7	27	0,040 [0,021-0,082]	83,36	1,37
R8	27	0,057 [0,028-0,100]	84,36	1,46
R9	27	0,058 [0,030-0,103]	82,95	1,66
P1	13	0,083 [0,039-0,164]	93,62	6,97
P2	11	0,015 [0,010-0,022]	91,21	7,52
P3	28	0,015 [0,010-0,022]	91,01	3,07
P4	28	0,024 [0,014-0,058]	86,58	9,72
P5	15	0,040 [0,020-0,060]	76,51	2,33
P6	28	0,022 [0,012-0,041]	93,86	5,23
P7	26	0,030 [0,021-0,046]	67,65	1,82
P8	28	0,012 [0,008-0,020]	78,19	1,65
P9	28	0,020 [0,011-0,032]	82,55	3,46
MW test Ref vs Pat	p=0,43	p=0,54	p=0,38	p<0,001

3 Results

The individual calculated parameters are given in Table 1. The subjects R1-R9 form the reference group (Ref), and P1-P9 are in the group of patients (Pat). The average speed is given as a median and quartiles because this parameter does not fit the normal distribution.

Despite the fact that all patients were Romberg's negative, three of them had the miniBESTtest scores from 11 to 15. Nevertheless, the difference between two groups (reference and patients) is not statistically significant.

The average speed and EFQ also have practically the same values in both groups, whereas the distance-based criterion *Id* substantially varies between groups. What is more important is that all these parameters are not correlated (Spearmen *R* absolute values are less than 0,33 with $p>0,15$).

The example of the raw data (*X* and *Y* coordinates of the selected point against time and its scattergram) is plotted in Figure 1. The displacements are within the 10 mm range, and scattergrams have no signs of imbalance. The cumulative percentages related to the EFQ index are shown in Figure 2. The cumulative distribution function of speed of the point was approximated by means of the empirical distribution step function.

The simplest way to explain the featured properties of our distance-based criterion *Id* is to represent the subjects, which involved in the study, as a set of point in the 3D space of per-coordinate components *Idx*, *Idy*, *Idz* (Figure 3). Note, that the scales of axis are different and the *Idx*, *Idy*, *Idz* are the square root components of *Id* in (3). It is clear that the reference group is quite compact, it members form one distinguishable cluster, while the patients are spread at a quite long distance from the reference group. This finding, on the one hand, highlights an inhomogeneity of the patient's group. On the other hand, the location of the certain patient may be considered either as an individual characteristic or as a pathological sign.

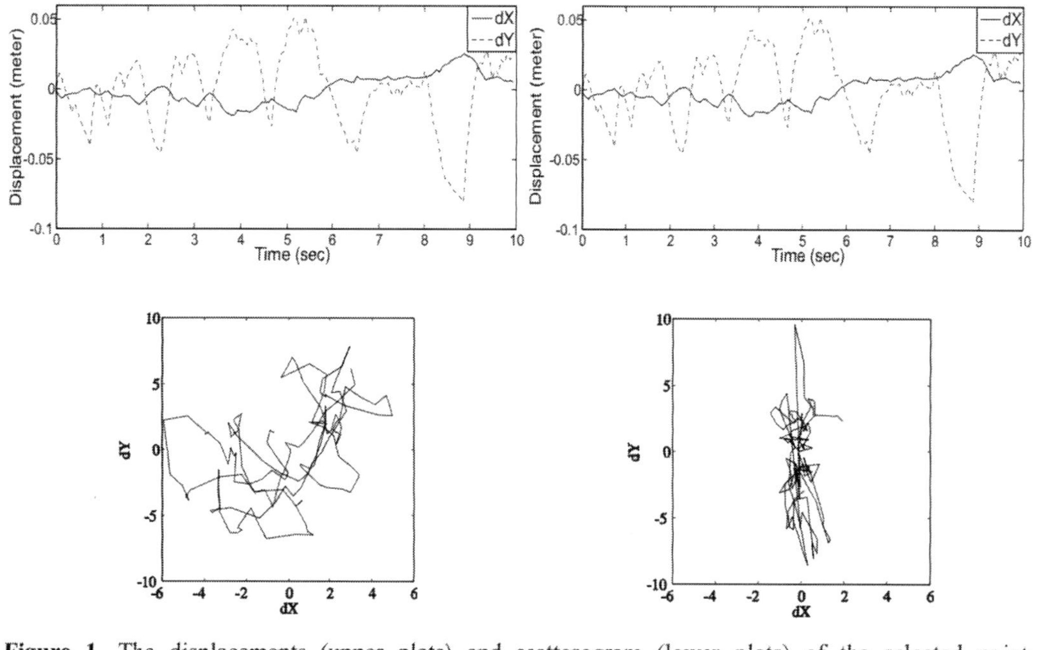

Figure 1. The displacements (upper plots) and scatterogram (lower plots) of the selected point for the healthy subject R2 (left plot) and patient P1 (right plot).

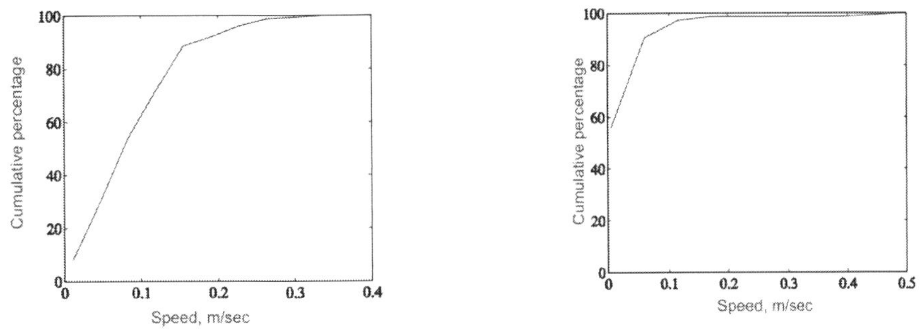

Figure 2. The cumulative percent of speed of the selected point for the healthy subject R2 (left plot) and patient P1 (right plot).

4 Discussion

The main result of our study is a simple and quite sensitive distance-based criterion to assess dissimilarity between, at least, two sets of measured values. The criterion mainly relays on the well-known Mahalanobis metric and inherited the ability to deal with correlated noisy data. This feature of the criterion is very important, because the key measurable parameters of the motion (positions and speed of the selected points) seem to be mutually linked. This may be a reason why some conventional methods, which imply statistical independence of the parameters, sometimes fail to estimate state of imbalance [8]. Another problem with these methods is that they are highly sensitive to noisy and inaccurate data. The distance-based approach has an innate ability to average chaotic signals and decrease noise proportionally to the square root of a vector's length [9]. We exploit this feature in a slightly different manner: the criterion requires a large reference group in order to achieve a desired confidence level, according to the formula (3). This feature may be considered as a drawback, but in practice of clinical studies a control (or reference) group of an appropriate size is an essential part of the research design. However, members of the reference group should be carefully selected.

The positive side-effect of the distance-based approach is that it reduces technical requirements to measuring devices. The accuracy and resolution of the motion capture hardware, which was used in our study, are sufficiently lower in comparison with the state-of-art force plates, but even in that case it was possible to distinguish between the groups.

5 Conclusion

A new approach to acquire and process data of postural tests has been proposed and proved to be sensitive under normal gravity conditions. The method takes into account small variations of the trajectory linked to individual physiological features of the vestibular system. A data gathering procedure does not involve the exact value of gravity, thus could be possible used under microgravity conditions. The main advantages of the approach are outstanding simplicity and low cost of hardware along with a high level of sensitivity. The specificity of the developed criterion towards the certain vestibular pathology is a discussable point and should be further investigated.

References

1. M. E. Pedalini, O. L. Cruz, R.S. Bittar, M.C. Lorenzi, S.S. Grasel, Acta Otolaryngol **129(9)**, 962 (2009)
2. N. Chastan, B. Debono, D. Maltete, J. Weber, Mov Disorders **23**, 366 (2008)
3. M. Mancini, F.B. Horak, E J Phys Med Rehabil **46(2)**, 239 (2010)

4. C. Casellato, M. Tagliabue, A. Pedrocchi, C. Papaxanthis, G. Ferrigno, T. Pozzo, Exp Brain Res **216**, 203 (2012)
5. M. Visser, J. Marinus, B.R. Bloem, H. Kisjes, B.M. van den Berg, J.J. van Hilten, Arch Phys Med Rehabil **84(11)**, 1669 (2003)
6. Computer stabiloanalyzer with biofeedback. URL: http://www.scenar.com.ru /production /stabile/index.htm (2016)
7. V.A. Fokin, Bull. TPU **311(5)**, 120 (2007)
8. R.J. Schilling, E.M. Bollt, G.D. Fulk, J.D. Skufca, A.F. Al-Ajlouni, C.J. Robinson, IEEE Trans. Biomed. Eng. **56(2)**, 292 (2009)
9. P. Nicolas, B-H. Avner, J. Classif **29(1)**, 23 (2012)

MATEC Web of Conferences **48**, 05002 (2016)
DOI: 10.1051/matecconf/20164805002

The possibility of distance methods application for snow dump sites monitoring

Olga Pasko[1,a], Elena Makartsova[1], Natalia Ushakova[1], Olga Tokareva[1] and Mikhail Mochalov[2]

[1] *National Research Tomsk Polytechnic University, 634050 Tomsk, Russia*
[2] *National Research Tomsk State University, 634050 Tomsk, Russia*

Abstract. In this article the results of remote sensing of the Earth for monitoring of four snow dump sites in Tomsk are described. Their compliance with permitted type of the territory use was evaluated. Earlier unknown time of the operation start was identified. The spatial-temporal variability of areas was defined. The temperature profiles of snow dump and background sites were analyzed. Use of remote sensing data allowed easy identification of snow dump sites creation time. The fact that the sites are located out of zones of permitted type of the territory use was revealed, that is violation of the law. For the first time the cartographic material was collected and showed that in the recent years their areas increased in average in 18%. The forecast for the nearest years was made. The article contains satellite images indicting the degradation of soil-vegetative cover of snow dumps. The reasons are contamination and overcooling of the soil in the beginning of vegetation period. The research results became the initial material for perfection of snow dumps territories management and will be applied in the work of environmental protecting service. Approaches proposed by authors may be used in solving similar problems in any region.

1 Introduction

Efficient management of land resources and making competent managerial decisions in many respects depends on completeness, quality and accessibility of information [1]. Traditional field researches as a rule are time-consuming, expensive, season and time limitations do not allow to obtain the necessary data of the land site conditions before the time of research.

The topical direction of territory conditions monitoring is application of Earth surface Remote Sensing Data (RSD) from the space which are more and more widely used in different fields of research [2]. It allows to define the boarders and square of land areas, their temperature, extent of degradation, environmental risks etc. not only at the moment but in the past time using the space images archive [3].

The objective of this work was revelation of RSD using ability for monitoring of snow dump sites in evaluation of compliance with permitted type of the territory use, spatial-temporal variability of areas and temperature profiles analysis.

The problem of snow dump sites is actual for the most part of Russian regions especially for Western Siberia due to the long period of snow precipitations. In particular in the territory of Tomsk

[a] Corresponding author : oap@tpu.ru

city the maximum of snow precipitations is from December until March, the maximum height of snow cover reaches 125 sm., and the duration of its occurrence is about half a year [4]. Snow is removed and accumulated on sites arranged for snow dumps. In the time of its melting the accumulated pollutants come to soils (possibly into underground water) and make environmental threat to the plants, animals and health of the population. It should be noted that Tomsk city and regional administrations do not have any information on borders and areas of snow dump sites and time of its operation start. It prevents to the definition of they age, pollutants accumulation level and necessary environmental safety works planning.

2 Data and methods of research

Objects of research were territories of four snow dump sites working in Tomsk situated in: Moskovskiy trakt 121 (provisionally CO1), Mostovaya street 40a (CO2), Khromovka settlement 35/2 (CO3), intersection of Ivanovskogo and Visotskogo streets (CO4).

The cartographical, spatial analysis of data and statistical methods of research were used.

For formation of snow dump sites borders and assessment of their areas changes form the time of creation until the current year the satellite images of high spatial resolution of Tomsk area obtained from cartographic service GoogleEarth for the period from 2006 till 2015 were used [5]. The territorial zones where the snow dump sites are situated were analyzed for compliance with permitted type of the territory use by the means of Tomsk urban atlas [6].

Evaluation of snow dump sites and background territories was made on the base of thermal data channel with spatial resolution of 100 m, obtained by the sensor TIRS (The Thermal Infrared Sensor) in the far infrared wavelength range 10.30 - 11.30 μm. Calibrated data of the mentioned channel provided by the USA Geological service (USGS) free of charge in the form of images with a radiometric resolution of 16 bits, were recalculated into temperature values in degrees Celsius [7].

Mapping and spatial analysis of data were made by the means of geoinformation systems MapInfo v8.5 and QGIS and software for the Earth remote sensing data processing Erdas imagine v9.2.

3 Results of research

The imposition of the coordinates of the snow dumps on the Urban Atlas map [6] allowed to define that snow dump site CO1 is situated in two territorial zones – social –business zone (sub-zone of business, public and commercial purpose) and recreational zone (sub-zone of urban parks, squares, gardens, boulevards). Snow dump sites CO4 and CO2 are situated in the production-business area (sub-zone of servicing facilities required for the implementation of production and business activities), and CO3 – in the recreational zone (sub-zone of city parks, squares, gardens, boulevards).

On the territory of the mentioned zones construction and expansion of operating industrial, communal and storage objects are not allowed except directly related to the operation of health and recreational, business, public and commercial objects [8]. Accordingly, there is a violation of the permitted types of land use of these ground areas which makes assessment of duration and negative impact level of snow dumps even more relevant.

Analysis of the satellite images in the period of 10 years allowed to exactly define that start of CO1 is 2013; CO2 – 2007; CO3-2010; CO4 – 2006. According to the existing images the borders of CO3 and CO4 for each year of work in the form of thematic layers of the digital map were constructed and they areas from the time of creation until 2015 were defined. On the figure 1, as an illustration, the images of snow dumps in the year of creation and in 2015 and their borders made on the base of these images was shown.

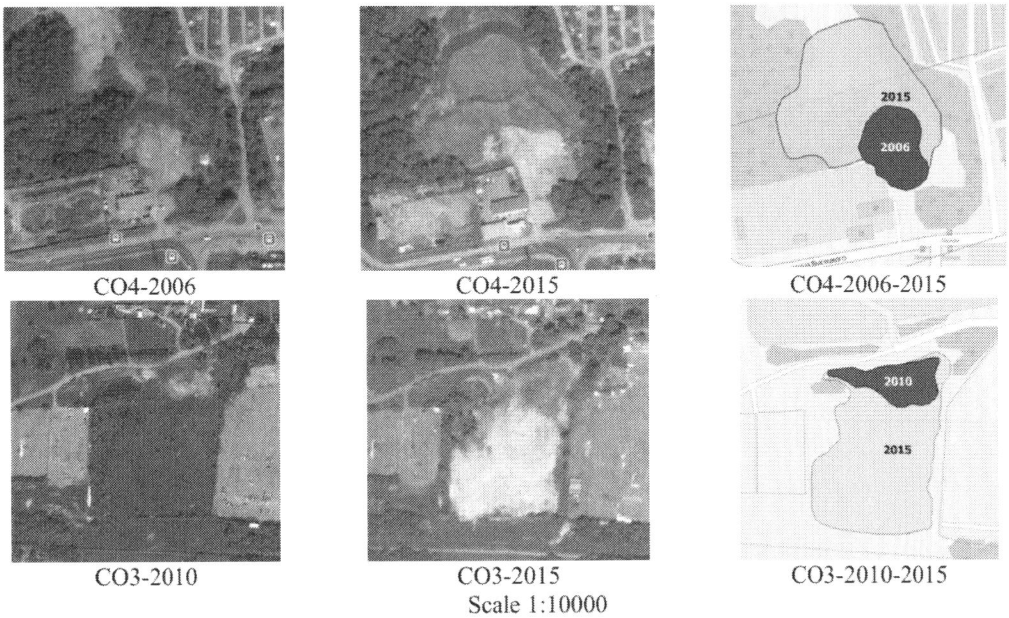

Figure 1. Satellite images and scheme of change of the snow dumps areas.

As it is shown in the figure 1 the territories of snow dumps are quite accurately decoded on the images of high spatial resolution indicated by degradation of vegetation cover especially in the autumn-winter period. Herewith the accuracy of tracing of the contour borders is very high and is situated in the limits of spatial resolution of the used image.

Analysis of dynamics of the snow dumps areas changes revealed a trend to increase of the area. Maximal growth was recorded for CO3 (+87 % to the initial area). An average dynamics of the snow dump sites growth in Tomsk for the 10 years period is 18%. Figure 2 shows a chart of snow dump CO3 change from 2009 till 2015 and forecast of development until 2017. It allows making a forecast for the nearest years. According to the chart the area CO3 increases approximately on 17 %.

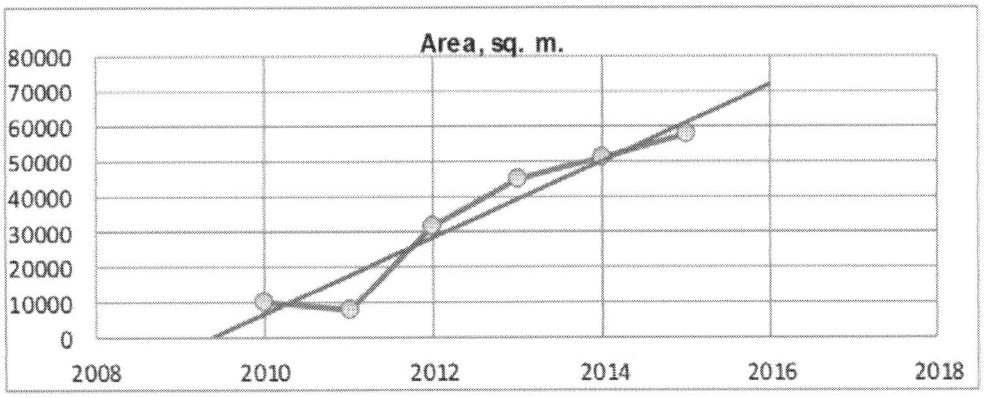

Figure 2. Forecast of the snow dump site CO3 area change.

Above it was stated that on the territory of snow dumps the degradation of vegetation which is not able to recover during the vegetation period was revealed. The most probable reasons may be:

- abnormally low temperatures of the soil in the beginning of vegetation period,
- excessive soil compaction by vehicles – dump trucks and bulldozers unloading and leveling of snow on a snow dump site;
- overwetting of the soil due to the long period of snow melting;
- soil contamination by the chemical reagents accumulated in the snow (exhaust gases etc.).

For analyses of the soil temperature changes the temperature profiles were traced for CO3 and nearby background site (control) – in the center of sites with the protrusion of 10 m. out of their borders from North to South. As it is shown on the figure 3, the character and dynamics of the profile temperature change on the snow dump and control site were significantly different. The temperature profile of the snow dump was characterized by the presence of an extremum. In the control it was absent. The difference between maximal and minimal temperatures in the frame of snow dump was in average 1,7° C, in control – 0,4° C. Approximately before the middle of May the central part of snow dump remained overcooled (on 2,5 ° C , after its heating happened, and to the middle of June temperature was higher than the control on 4,3 ° C, to the end - on 5 ° C). Inhibition of germination of plants by low temperatures in early stages affected their growth and development, which is planned to study in the future researches.

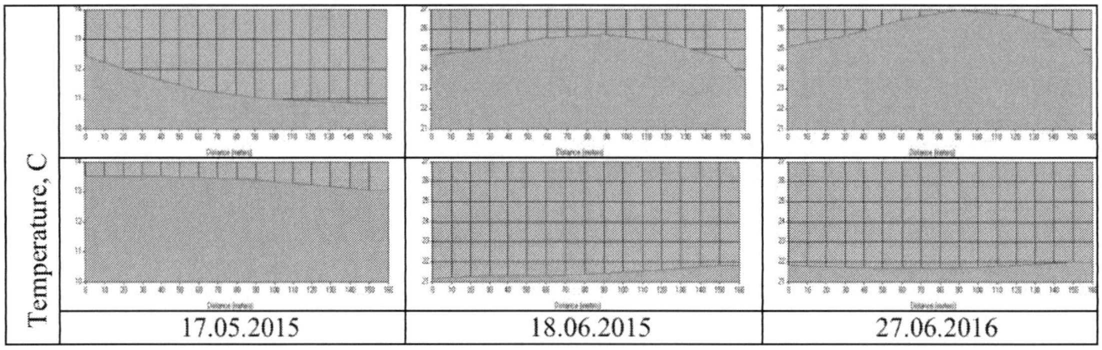

Figure 3. Temperature profiles for CO3 (upper row) and background site (the lower row) in May-June 2016.

4 Conclusion

The use of RSD allowed to obtain actual information about snow dump sites in Tomsk which was not possible to do by any other methods:

- allocation of snow dumps out of areas permitted for this type of the land use was defined;
- for the first time the time of snow dumps creation in Tomsk was revealed;
- the cartographic material, reflecting dynamics of snow dumps borders was collected;
- the trends of the snow dump sites' areas changes was analyzed and forecast of their changes in the nearest years was made;
- abnormal temperature regime on the snow dump sites was revealed.

The results obtained during the work may be initial material for perfection of snow dump site territory management and be applied in monitoring of environmental services for strengthening of environmental safety of the territory.

References

1. O. Pasko, T. Mochalova, IOP Conf. Ser.: Earth Environ. Sci. **21**, 12044 (2014)
2. L. Altunina, L. Svarovskaya, Y. Polishchuk, O. Tokareva, Petr. Chem. **51**, 381 (2011)
3. L. Feng, X. Han, C. Hu, X. Chen, Rem. Sens. Environment. **176**, 43 (2016)
4. Weath. & Clim. Electronic resource. URL: http://www.pogodaiklimat.ru/climate/29430.htm (2016)
5. GoogleEarth. Planet Earth for PC. Electronic resource. URL:http://www.google.com/earth/ (2016)
6. Urban atlas of Tomsk. Electronic resource. URL: http://map.admin.tomsk.ru/ (2016)
7. U.S. Geological Service. Electronic resource. URL: http://landsat.usgs.gov/landsat8.php (2016)
8. Urban zoning. Electronic resource. URL: http://www.zemvopros.ru/page_4603.htm (2016)

MATEC Web of Conferences 48, 05003 (2016)
DOI: 10.1051/matecconf/20164805003

Using MODIS NDVI products for vegetation state monitoring on the oil production territory in Western Siberia

Anton Kovalev[1,a] and Olga Tokareva[1]

[1] *National Research Tomsk Polytechnic University, 634050 Tomsk, Russia*

Abstract. Article describes the results of using remote sensing data for vegetation state monitoring on the oil field territories in Western Siberia. We used MODIS data product providing the normalized difference vegetation index (NDVI) values. Average NDVI values of each studied area were calculated for the period from 2010 to 2015 with one year interval for June, July and August. Analysis was carried out via an open tool of geographic information system QGIS used for spatial analysis and calculation of statistical parameters within chosen polygons. Results are presented in graphs showing the variation of NDVI for each study area and explaining the changes in trend lines for each field. It is shown that the majority of graphs are similar in shape which is caused by similar weather conditions. To confirm these results, we have conducted data analysis including temperature conditions and information about the accidents for each area. Abnormal changes in NDVI values revealed an emergency situation on the Priobskoe oil field caused by the flood in 2015. To sum up, the research results show that vegetation of studied areas is in a sufficiently stable state.

1 Introduction

Western Siberia is the largest oil and gas basin in 3.5 million kilometers square in space and is one of the main areas for oil and gas production in Russia. Oil and gas industry activity leads to systematic damage to the vegetation as a result of mechanical damage to the surface at the stage of research and infrastructure development of oil and gas producing fields. This damage occurs due to fires, oil spills, oil drilling fluids, and the air pollution impact on vegetation. Therefore, permanent environment monitoring is required on the territories under the constant intensive technological pressure.

The chemical impact of oil and gas industry on environment is caused by toxicity of natural hydrocarbons and chemical substances used in manufacturing processes. Natural recovery processes of the vegetation cover after accidental spills exposure are fairly long so natural components need restoration and reclamation [1]. The problem of accidental oil and various reagents spills is often solved by dumping sand in violation of technological requirements. In these cases the problem is only complicated, because the pollutants remain in the soil and fall into the surface and underground waters.

Gas flares scattered throughout the West Siberian oil and gas province produce very aggressive and dangerous chemical pollution agents [1]. This problem is especially relevant for Khanty-Mansiysk autonomous okrug (KMAO) located in the area of the southern, middle and northern taiga dominated by

[a] Corresponding author : Optimist93@sibmail.com

coniferous forest particularly sensitive to pollutions [1]. In addition, there are signs of an active pest infestation in the autumn period which is caused by the heating effect of the flares and the subsequent weakened state of the trees.

Nowadays, satellite monitoring data is used in addition to traditional methods of ground-based observations of vegetation cover [2, 3]. The remote sensing data allows monitoring of the state and structure of vegetation over large areas, including those hardly accessible for ground survey due to high water logging. In the paper [3] authors discuss the results of using NDVI [2, 4], which was calculated with Lansat data (spatial resolution 30m) to evaluate the negative impact of chemical pollution on the state of coniferous trees in the flares' area of influence at Priobskoe oil field. It was shown that the results of remote assessment of the forest vegetation state changes based on vegetation indices sufficiently correspond to these bioindicative ground research results.

The purpose of this study is to evaluate the changes of vegetation in the oil production field territory on the basis of NDVI values using data obtained by the MODIS (Moderate Resolution Imaging Spectroradiometer) instrument [2] with spatial resolution 500 m.

2 Study areas

Research was conducted in five oil-producing areas: Khanty-Mansiysk autonomous okrug (Ust-Balykskoe, Priobskoe, Samotlorskoe and Lyantorskoe oil fields) and in Tomsk region (the Vasuganskaya group of oil fields: Pervomaiskoe, Lomovoe, Olenie, Katylginskoe, Lontynyahskoe). A segment of the Ugansky Park was chosen as a background area. The boundaries of the oil field territories were constructed using Landsat-8 satellite images with a spatial resolution of 30 m, which allows deciphering the field infrastructure (holes, pipelines, roads, etc.). Figure 1 displays a scheme of studied areas shown as polygons of a thematic vector layer on a digital map, where boundaries of the segments are shown in thick black lines.

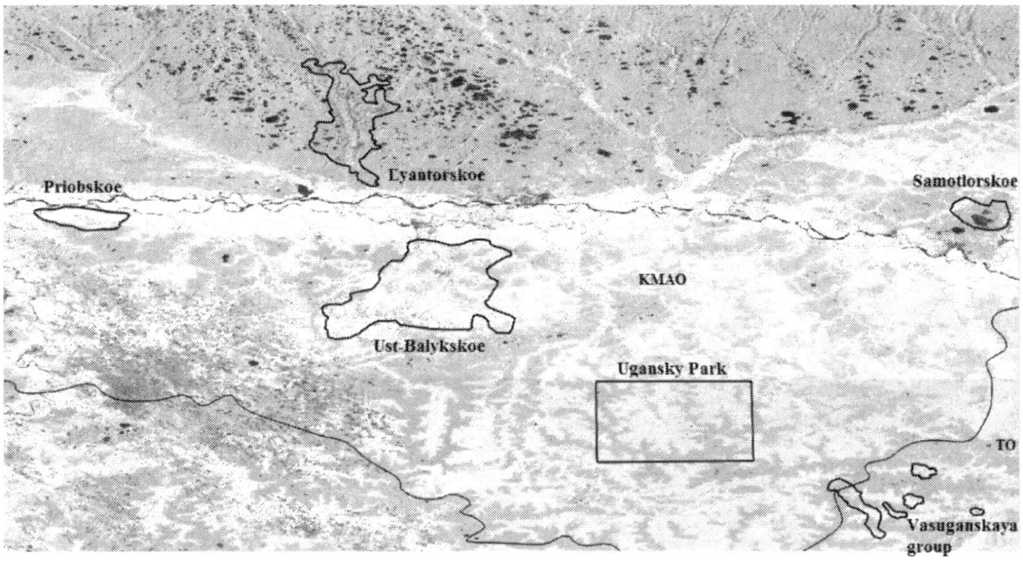

Figure 1. The location scheme of studied areas. Legend: TO – Tomskaya Oblast.

3 Data and research methods

NDVI has been chosen for this analysis as it has high sensitivity to biomass changes and is suitable for quantitative measure. It has fixed values for objects not related to vegetation: the value 0.7 corresponds to luxuriant vegetation, 0.5 – sparse vegetation, 0.025 – soil, 0 – clouds, -0.05 – snow and ice,-0.25 – water, -0.5 – man-made materials (e.g. concrete, pavement). According to the evaluation of given values from images, a structure of an explored area and vegetation may be considered [4]. Below is the formula of calculation for the index:

$$\mathrm{NDVI} = \frac{Pnir - Pred}{Pnir + Pred}, \tag{1}$$

where *Pnir* – spectral luminance pixel values in the near infrared range of the electromagnetic spectrum; *Pred* – in the red range of the spectrum.

Study used MODIS MOD13Q1 Vegetation Indices 16-Day Global 500m with 16-days average NDVI values provided by an online archive of NASA EOSDIS. For the purpose of correct data visualization on the screen NDVI values in provided files are stored with the scale coefficient of 10000. According to the MODIS scheme of a segments location, studied areas are shown on images using numbers (21, 02) and (21, 03).

Spatial data analysis was selected as the main research method to evaluate characteristics of the region of interest. Spatial analysis was conducted using QGIS tools created to calculate the statistical indicators within the selected polygons.

4 Results of research

To study the dynamics of changes in vegetation state within each area, average values of NDVI by year were calculated for the period from 2010 to 2015, with an interval of one year. We used data, obtained from the pictures with recording dates 12.06-27.06, 12.07-27.07. Average NDVI values and trend lines for each study area are shown in Figures 2 and 3. They show that behavior of NDVI changes for each studied area. It is worth noting that NDVI values of the Lyantorskoe and Samotlorskoe oil fields significantly differ from values of other areas (approximately 0.1 less). Such a variation is caused by a lot of swamps within the total area of the field (about 1000 sq km for Samotlorskoe). As mentioned above, the NDVI value for water is low.

As we can see from Fig. 2, the majority of graphs have similar shapes caused by weather conditions. Analysis of temperature data showed that in 2013 the spring was late and at the beginning of June the weather was cold. It caused delay in building biomass plants at the beginning of vegetation period and explained the reason why NDVI values were low in July 2013 (Fig. 2).

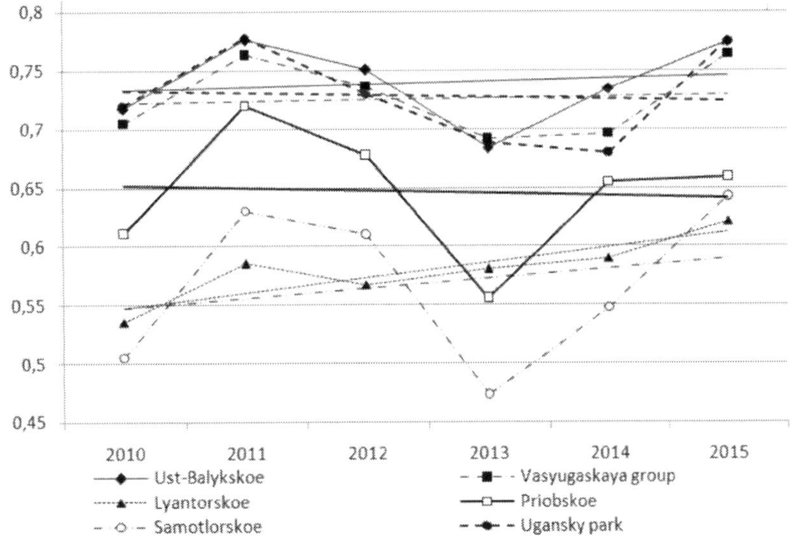

Figure 2. Changes in average NDVI values by year (recording dates 12.06-27.06).

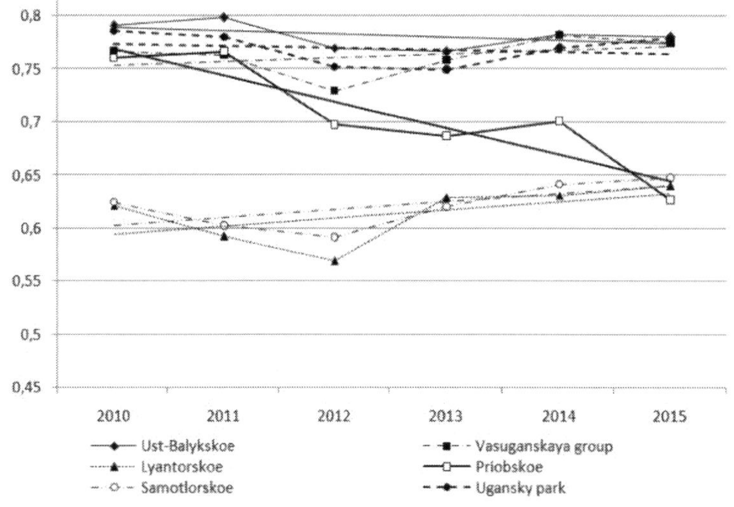

Figure 3. Changes in average NDVI values by year (recording dates 12.07-27.07).

Dry and extremely hot weather in June and July 2012 caused significant reduction of NDVI values in the second half of July in each study area. Slope of the trend lines is not significantly different for the territories of the Ugansky park, Ust-Balykskoe, Priobskoeoil fields and Vasuganskaya group of oil fields. The trend angle of the Lyantorskoe and Samotlorskoe oil fields shows an improvement in the state of vegetation that can be associated with the ongoing remediation work here. At the moment the oil production on the Lyantorskoe oil field is decreased. That causes self-healing processes of land and, therefore, vegetation.

Comparative analysis of trends in Figures 2 and 3 is shown that NDVI values are slightly increasing in July. This fact associated with changes of vegetation period, but in some parts the NDVI values do not change or even decline in July of the same year. For example, in July 2015 there is a significant decrease in NDVI values compared with the values obtained in June on the Priobskoe oil field. NDVI changes on other oil fields are slight. This is explained by a high level of flood waters in July 2015 and a large amount of precipitation. A significant part of territory of the Priobskoe oil field is located in the overflow land of the Ob river. At the end of June 2015 an accident occurred on the oil pipe line on the territory of the Ust-Balykskoe oil field, but the bulk of the oil got into the river.

5 Conclusion

We can conclude that vegetation state in the study areas is sufficiently stable. Thermal and water regime has the greatest influence on the average NDVI values here. Comparative analysis of graphs of NDVI values reveals abnormal changes in the territory and determines areas which require a more detailed analysis of the reasons using the ground-truth data and satellite images with high spatial resolution.

References

1. L. K. Altunina, L. I. Svarovskaya, Y. M. Polishchuk, O. S. Tokareva, Petr. Chem. **51 (5)**, 381 (2011)
2. T. Hilker, A. I. Lyapustin, F. G. Hall, R. Myneni, Y. Knyazikhin, Y. Yang, Y. Wang, C. J. Tucker, P. J. Sellers, Rem. Sens. of Env. **166**, 233 (2015)
3. O. S. Tokareva, Y. M. Polishchuk, Sovr. Prob. Dist. Zond. Zemli iz Kosm. **10 (2)**, 81 (2013)
4. J. R. Jensen, *Remote Sensing of the Environment: At Earth Resource Perspective* (Pearson Education, 2011)

MATEC Web of Conferences 48, 05004 (2016)
DOI: 10.1051/matecconf/20164805004

Development of a high-resolution apparatus to monitor physiological state of a person undergoing extreme conditions

Radik Kodermyatov[1,a], Maxim Ivanov[1], Mikhail Yuzhakov[1], Vladimir Kuznetsov[2], Mariya Yuzhakova[1] and Evgeniya Timofeeva[1]

[1] *National Research Tomsk Polytechnic University, 634050 Tomsk, Russia*
[2] *Omsk State Technical University, 644050 Omsk, Russia*

Abstract. The present research has shown that ECG recordings and parameters of the body functional state are crucial for all kinds of the astronaut pre-flight preparations (centrifuge, thermal chamber, pressure chamber, pressure chamber with special equipment). It is, therefore, important to develop methods and tools for early detection of the preclinical forms of the functional state disorders in patients undergoing high-intensity loads of mixed character. The method based on the hardware-software complex (HSC) with nanosensors of high resolution has been proposed to measure the electrophysiological characteristics and bioelectrical impedance of the body tissues directly exposed to loading. The hardware-software complex is subjected to clinical trials in Tomsk Research Institute for Cardiology. The obtained results show that the use of nanosensors of high resolution in the HSC without standard filters allows elimination of the power-line interference in ECG recordings. Monitoring of the tissue bioimpedance parameters under test loads enables the detection of preclinical (latent) forms of various diseases. After clinical trials in Tomsk Institute for Cardiology the developed apparatus will be proposed for in-depth study of the cardiovascular system and the functional state of the body of astronauts and other persons exposed to extreme loads.

1 Introduction

The human body affected by various environmental factors attempts to compensate the impact of the factors, which can cause abnormalities in its organs and systems. This is due to intensification of the entire body and its tissue responses to external forces, for instance, violent changes in air pressure cause the compensatory hyperactivity of cardiovascular system to restore the state of homeostasis, that in turn leads to additional loads on other body systems – circulatory, musculoskeletal, respiratory, nervous, endocrine, etc. These processes can lead to some abnormalities in living tissues causing the resistant forms of chronic pathologies and acute forms of disorders in the organism functional state.

Early detection and prevention of abnormalities are critical for the extreme professional activities related to a long-term presence in isolated environments such as deep underwater and underground work, space flights, and others.

[a] Corresponding author : kodermyatovra@mail.ru

The present study proposes the method based on the hardware-software complex with highly sensitive nanosensors to record biopotentials and bioelectrical impedance of living tissues in individuals exposed to extreme environments. The method allows recording abnormalities in the functional state of the tissues with high accuracy.

The study aims to investigate the effect of extreme environment on the functional activity of the human organs and tissues.

The objectives of the study are as follows: to consider the effect of extreme loads on electrophysiological characteristics of the activity of cardiovascular and musculoskeletal systems in healthy individuals; to compare the cardiovascular system parameters and the functional state of particular tissues in healthy individuals under extreme working conditions with those of having some pathologies; to propose a reliable method to monitor physiological state of individuals exposed to loads of mixed character and high intensity at the stage of initial selection, preliminary tests and simulation of extreme working conditions.

2 Methods and research equipment

The research equipment used in the study is a hardware-software complex with highly sensitive components for a three-lead ECG recording in the frequency range from 0 to 10000 Hz with the measured signals ranged from ± 0.3 μV to ± 10 μV and bioimpedance tissue recordings in the frequency range from 10 to 100000 Hz with the range of the impedance module measurement from 0.3 to 1000 kΩ.

Those whose professional activity relates to the long-term presence in extreme environments, for instance, astronauts or deep divers, must undergo a number of stages of health monitoring, both in the preparatory period and before and during the performance of professional duties [1]. It is obvious that health assessment is crucial in the recreation and rehabilitation (in the event of temporary disability) periods due to the possible occurrence of hidden health problems in case of the medical care unavailability.

Thus, the complete and reliable diagnostic information on possible occurrence of health disorders ensure a high probability of timely preventive, therapeutic and organizational measures and, therefore, minimize the risk of emergencies in the isolated unfavorable environment.

In space medicine and labour protection, many factors affect a person working in extreme environments [2]. Most of the factors are associated with the physical effects on the body – vibroacoustic, light, electromagnetic fields, ionizing and non-ionizing radiation, air environment characteristics, microclimate, and others. The combined effect of these factors, especially their extreme values, on the body triggers its reflex mobilization to preserve the state close to a healthy homeostasis, which in turn leads to the increase in loads on the corresponding systems – nervous, cardiovascular, musculoskeletal, endocrine, and others.

A significant factor of extreme physical effects on humans in spaceflight is the so-called "overload" which indicates the excess of the action of mechanical forces over the body weight. Due to the vector nature of this magnitude, overloads can be positive (action of forces is directed vertically upward from the Earth's surface) and negative (action of forces is directed towards the Earth's surface and the internal organs are "pulled down" to the opposite side regardless of the overload direction).

The overload may be long acting or shock depending on the time and intensity of exposure. The shock load occurs when the capsule is ejected from the mother ship; long-term load takes place when the spacecraft starts orbiting the Earth.

The astronauts experience a long-term load at the liftoff and descent of the spacecraft (Figure 1). Special centrifuges are used to simulate this kind of load [3, 4, 5].

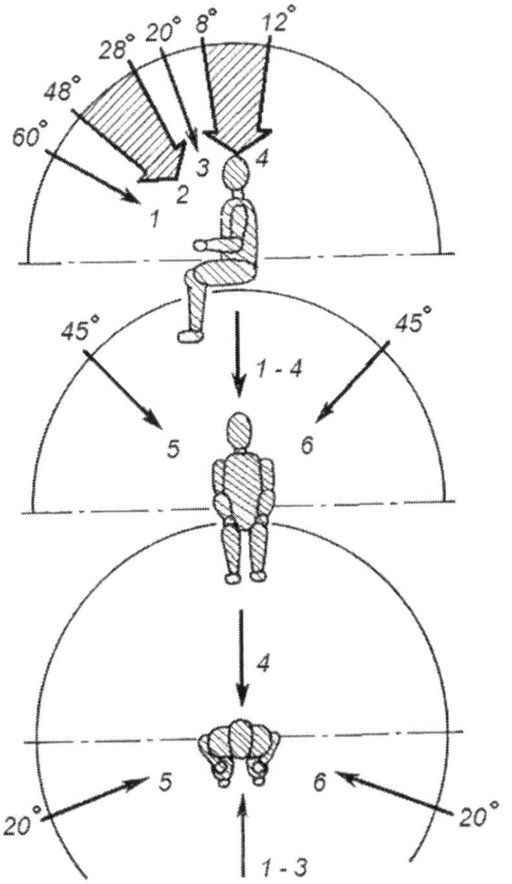

Figure 1. Schematic diagram of the overload direction relative to a seated person.

Weightless environment affects the body when the spacecraft starts orbiting the Earth and the gravity is balanced by inertial forces or almost disappears. The effect of weightlessness on the cardiovascular system results in the heart rate increase with periodical increase in pulse rate and blood pressure lowering.

The body response to weightlessness appears in sharp intensification of cardiovascular activity and hyper-compensatory reactions of muscle tissues [3, 4, 5].

The physiological distress in case of chronic disease or other latent diseases may cause the symptom manifestation, which in turn can lead to great difficulty in providing emergency medical care in the isolated environment. Effective diagnostics at the stages of preflight preparation and assessment of the astronauts' health in the course of their professional activity will enable elimination of numerous problems associated with a "human factor" in the flight program implementation.

At the stages of preflight preparation (general space training (GST), specialized space training (SST) and preflight space training (PFST)), which total duration is from 26 to 52 weeks, general health, physical fitness, capability to endure the exposure to space environment, overload, weightlessness and other parameters are constantly monitored.

The recordings of cardiovascular system parameters are critical for all types of astronaut testing (centrifuge, thermal chamber, pressure chamber, pressure chamber with special equipment); the parameters of musculoskeletal, nervous and respiratory systems are essential for some other methods.

The obtained information allows not only to assess the overall state of the body, but also to determine the level of preparedness and capability to endure the impact of extreme environments.

Therefore, there is an objective need to develop methods and tools for early detection of preclinical forms of diseases in individuals experiencing stresses and weightlessness. The cardiovascular diseases and diseases of spine are the most common causes for the astronaut disqualification on medical grounds. The electrophysiological methods are primarily used for the selection and examination of astronauts (electrocardiography (ECG) at rest, rheoencephalography (REG), electroencephalography (EEG), veloergometry with the dosed physical load, electromyography (EMG)).

The research group of Medical Instrument-Making Lab No. 63, Tomsk Polytechnic University, in collaboration with specialists of Omsk State Technical University have developed a hardware-software complex with highly sensitive elements to record a three-lead ECG in the frequency range from 0 to 10000 Hz with the measured signals ranged from $\pm$ 0.3 μV to $\pm$ 10 μV, and to record the tissue bioimpedance in the frequency range from 10 to 100000 Hz and the impedance module measurement range from 0.3 to 1000 kΩ.

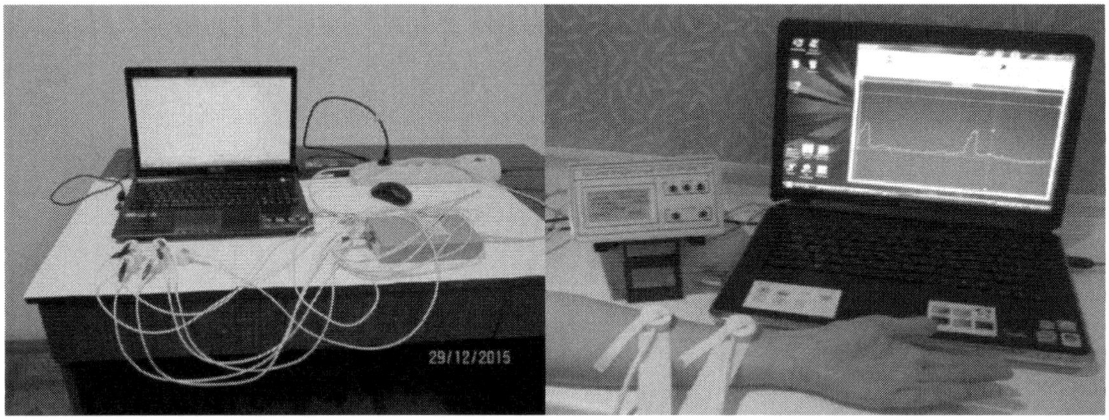

Figure 2. General view of the HSC elements (the ECG apparatus is on the left, the apparatus for bioimpedansometry is on the right).

The following are the results of the proposed HSC implementation for diagnostic studies performed in the Department of Emergency Cardiology, Tomsk Research Institute for Cardiology.

3 Results

A three-lead ECG measurements are regularly carried out (the frequency of medical examination is determined by the medical monitoring requirements for individuals who are allowed to work in extreme environments) without filters typical of conventional electrocardiographic equipment (high frequency filters, notch filters and tremor filters). Figure 3a, b shows a high-resolution ECG and its spectrum recorded in a postinfarction patient in the emergency department of Tomsk Institute for Cardiology.

As can be seen from the ECG obtained without filtering, power-line interference is not found in the recorded ECG that allows reliable detection of minor changes in the patient's cardiovascular system at the early stages of pathology.

Measurements of the tissue bioimpedance are performed using the two-electrode non-invasive method by placing the electrodes on certain areas of the patient's tissue (for example, motional groups of the upper and lower limbs). The measurements are carried out in three stages: before, after and an hour after the test load on the musculoskeletal system; the results are compared with the parameters of the objective state of a person obtained by other methods. The information on the dynamics of changes is also analyzed [7].

Figure 4 shows the approximated spectrograms of the musculocutaneous tissue of the lower extremities in patients with different levels of physical fitness (red curve indicates the change of the impedance module and green curve indicates a phase angle): a) Group 1 (high level of physical fitness); b) Group 2 (basic level of physical fitness); c) Group 3 (unsatisfactory level of physical fitness).

The bioimpedance analysis has shown that the spectrogram of individuals with a high level of physical fitness and capable of enduring extreme environments over a long period of time differs from the spectrogram of individuals with unsatisfactory level of physical fitness in the entire frequency range (Figure 4).

a)

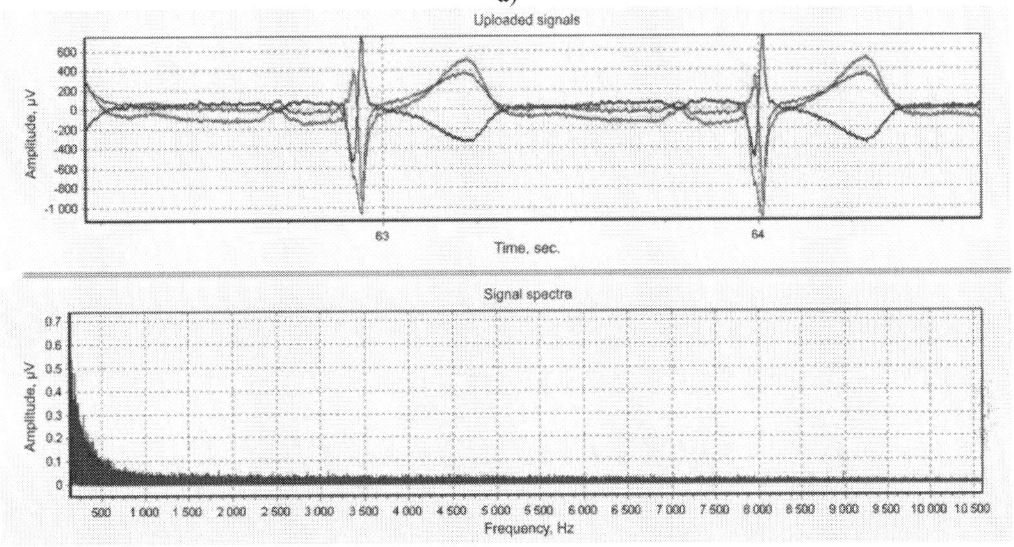

b)

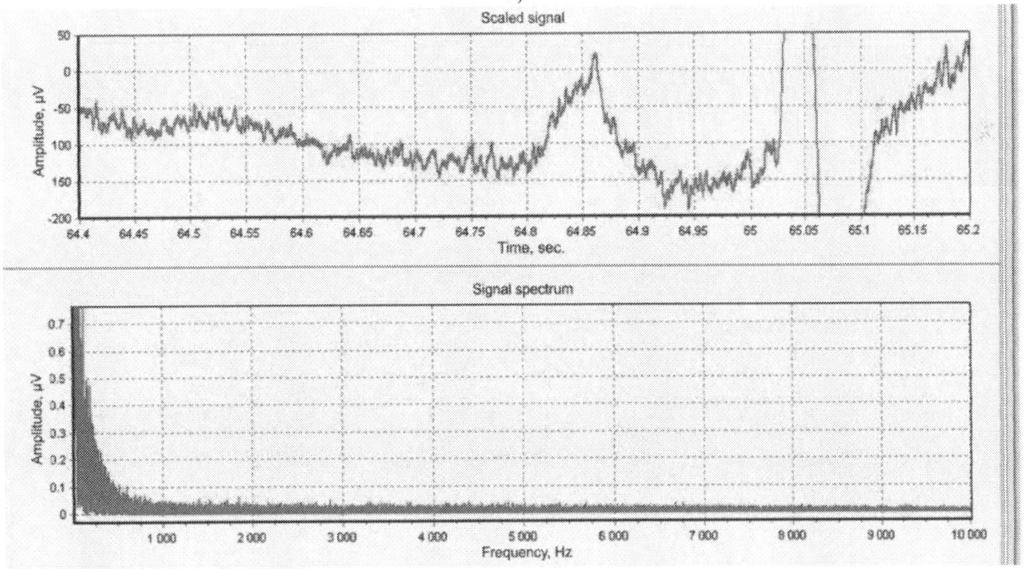

Figure 3. Patient's ECG: a) a three-lead ECG recorded from the patient's chest by the Holter method: 1 (red) – 1 lead, 2 (green) – 2 lead, 3 (blue) – 3 lead, b) ECG fragment of the 1st channel (1 lead) with the heart micropotentials at the scale sensitivity of 50 µV/div .

The examined individuals with no visible abnormalities in the cardiovascular system on the obtained ECG recordings experienced changes of the impedance module and its phase angle. The changes were similar to those of healthy people with a high level of physical fitness [8].

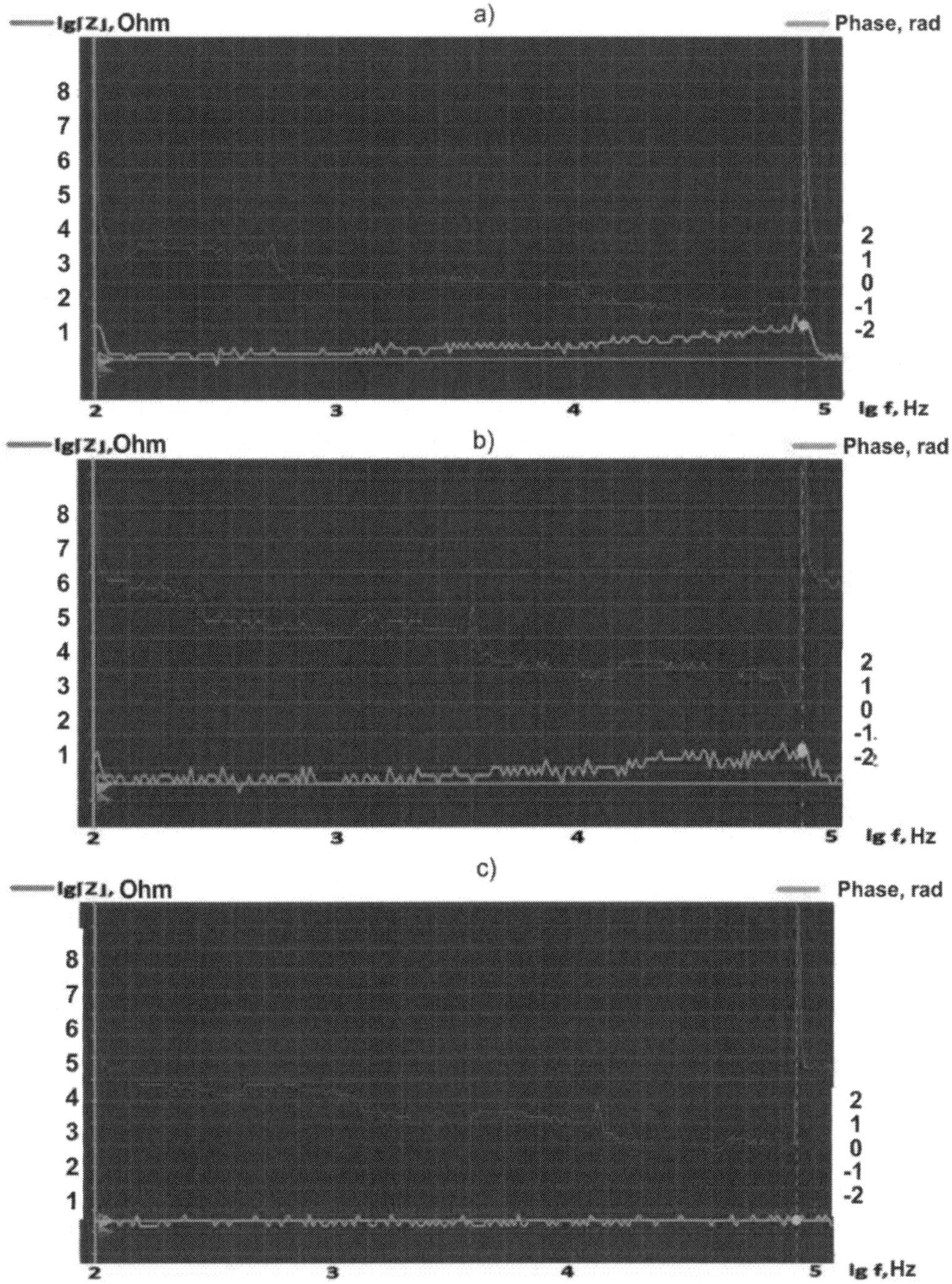

Figure 4. Approximated spectrograms of the musculocutaneous tissue of the lower extremities in patients with different levels of physical fitness (red curve indicates the change of the impedance module and green curve indicates a phase angle): a) group 1 (high level of physical fitness); b) group 2 (basic level of physical fitness); c) group 3 (unsatisfactory level of physical fitness).

4 Conclusion

The present study has shown that the diagnostic information on the functional state of the body tissues and the cardiovascular system activity obtained by the developed HSC makes it possible to assess the person's capability to endure high-intensity loads of mixed character typical of space missions or underwater work. The diagnostic information helps to identify possible pathologies induced by the loads.

Further research aims to early detect the minor changes in the person's cardiovascular and musculoskeletal systems, which cannot be recorded by the common methods without highly sensitive means of obtaining diagnostic information.

After clinical trials in Tomsk Institute for Cardiology, the developed apparatus will be proposed for in-depth study of the cardiovascular system and functional state of the body of astronauts and other persons undergoing high-intensity loads of mixed character both for primary selection of astronauts and assessment of their health in the course of the professional activity.

Acknowledgement

The research was financially supported by the Federal Targeted Programme "Research and Development in Priority Fields of S&T Complex of Russia in 2014–2020", the Agreement No. 14.578.21.0032 dated 05.06.2014 "Development of the experimental sample of a hardware-software complex for noninvasive recording of heart micropotentials in a wide frequency band without filtering and averaging in real time to early detect the symptoms of sudden cardiac death", the unique identifier of the contract: RFMEF157814X0032.

References

1. Y. Y. Xue, G. Y. Liu, B. S. Xie, X. Y. Chen, Y. Cheng, Space medicine & medical engineering **13**, 104 (2000)
2. GOST 12.0.003-74 Occupational safety standards system. Dangerous and harmful production effects. Classification
3. A. I. Grigoriev, A. R. Kotovskaya, G. A. Fomina, Acta Astronautica **68,** 1495 (2011)
4. G. M. Pantalos, T. E. Bennett, M. K. Sharp, S. J. Woodruff, S. D. O'Leary, K. J. Gillars, T. Schurfranz, S. D. Everett, M. Lemon, Current Pharmaceutical Biotechnology **6,** 331 (2005)
5. S. Ilyas, M. Aslam, Medical Forum Monthly **13,** 1 (2002)
6. Kriterii otbora v kosmonavty i kosmicheskie turisty. Spravka. URL: http://ria.ru/spravka/20100428/227827425.html (2016)
7. V. V. Kuznetsov, A. A. Novikov, Omsk Scientific Bulletin **2(116)**, 235 (2013)
8. V. V. Kuznetsov, A. A. Novikov, VI All-Russian Scientific Conference "Young Russia: advanced technology – in the industry" **3**, 229 (2015)

MATEC Web of Conferences **48**, 05005 (2016)
DOI: 10.1051/matecconf/20164805005

Sensory Dissociation in Vestibular Function Assessment

Ivan Tolmachev[1,a], Konstantin Brazovsky[2], Sergey Schadenko[2] and Ekaterina Korzhenkova[1]

[1] *National Research Tomsk Polytechnic University, 634050 Tomsk, Russia*
[2] *Siberian state medical university, 634050 Tomsk, Russia*

Abstract. The project aims to create a solution to the problem of early diagnostics of neurodegenerative disorders, accompanied by imbalance, and to develop rehabilitation methods for patients with vestibular disorders. The balanced upright position is a result of interaction between vestibular system, skeleton and muscles, visual, and proprioceptive systems. Postural abnormalities are caused by morphological or functional alteration of the components of the equilibrioception system. These disorders weaken the ability of the central nervous system to process vestibular, visual and proprioceptive signals, which are responsible for the sense of balance. Moreover, it also decreases the plasticity of the nervous system, resulting in adaptation disfunctionality. Unfortunately, premorbid detection of vestibular dysfunctions is quite a complicated test because available methods to evaluate postural reflexes have low sensitivity and specifity. Consequently, the development of methods to assess functional state of the equilibrioception system becoming viable in order to detect neurodegenerative disorders as early as possible, to control treatment and rehabilitation procedures.

1 Introduction

Application of sophisticated computer methods for medical diagnostics, as well as for scientific research and education became the common practice these days. Modern automated complexes allow obtaining research data quickly and with a high confidence level. Unfortunately, increase of the volume of information to process doesn't lead, automatically, towards boost of diagnostics quality and precision. The point is that the obtained raw data characterizing the state of the complex multilevel biological system at the specified moment of time. However, the system state itself is not the essential indicator of biological system ability to adapt and function normally. We are proposing the convenient approach towards correlation of measured parameters of biosystem with the needed values. The approach is based on integrated evaluation of condition and allows to consolidate heterogeneous multidimensional data into the integrated criteria, which characterizing the rate of system deviation from some reference state (e.g. norm, average value, etc.). It was shown that the proposed approach could be applied for data of different nature (e.g. infrared spectra of absorption spectrum of exhaled air, tomographic images, electro-physiological measurements, etc.). The only limitation of the approach is the necessity to have powerful computational resources, to model the states of biosystems using varied set of data. However, the problem could be solved by utilizing the parallel and high-performance computing. During the analysis of dynamic processes in biology and medicine in general, and in the diagnostics of human movement, in particular, the large amount of data is

[a] Corresponding author : ivantolm@mail.ru

generated. Posturology is a relatively new trend in neurology. This branch focuses on the development and use of various assessment methods and approaches for the diagnostics of disorders of human balance system. One method to evaluate the functional state of the vestibular system is to record the behaviour of the overall pressure centre on the surface with a force platform. This method, called stabilometry, is actually the "gold standard" in postural studies. Nevertheless, it has several drawbacks, for example, fairly large dimensions of the equipment that must be installed properly and carefully calibrated before use. Usage of the high-performance computing environment allows to create a multi-parametric evaluation system of movement quality and functional equilibrium state of a human, which includes a set of techniques, such as dynamic stabilometry, video capturing and detection of movements, electroencephalography, oculography, myography. Meanwhile, in recent years, due to the development of computer graphics, we have an opportunity to create a virtual three-dimensional visual environment, which simulates the actual environment. The virtual environment could be easily controlled via a computer to simulate the desired motion and thus to carry out detailed studies of various aspects of visual information in the postural control. Postural control of the patient without vestibular disorders is realized through the visual signals. The virtual environments can cause a conflict between the perception of movement, created by the visual system with the vestibular and proprioceptive input signals, indicating a static state. This vestibular-visual and proprioceptive conflict of the "neural redundancy" is weakened or even completely absent, if there is an acute sensory deprivation, which affects postural control. We assume that the postural studies during the process of specific impact of the virtual environment could allow assessing postural strategy and the role of each signal stream: visual, vestibular and proprioceptive. In terms of diagnostics, such approaches can detect the process of adaptation of the human balance system, hence allows detecting the presence of a compensatory process at an early stage, enhancing the function of the visual analyzer in maintaining the vertical position of the person.

Postural disorders may be connected with diseases of various organs and tissues, which affect the normal functioning and determine the clinical manifestations and prognosis. Causes may be different: injuries, birth defects and predispositions, influence of food, environmental and genetic factors, or idiopathic reasons [1]. All these disorders affect the central and the peripheral nervous system, and the musculoskeletal system, disrupting the balance and posture of the individual [2].

One of the most frequent causes of balance disorders and postural instability is Parkinson's disease [3]. Parkinson's disease is a neurodegenerative disease that affects the central nervous system [4, 5, 6], and is clinically characterized by slowness of movement, resting tremor, rigidity, postural disorders and a number of non-motor symptoms. The etiology of this disease is still unknown; however, the main cause of Parkinson's disease is the loss of dopaminergic cells primarily in the nigrostriatal system [7], which leads to movement disorders.

Parkinson's disease is a progressive disorder, and during the disease the motor defect becomes more pronounced, which explains the emergence and intensification of symptoms such as resting tremor, rigidity, bradykinesia, changes in posture, instability, impaired balance and walk [8]. Moreover, this disease decreases the ability of the central nervous system to process vestibular, visual and proprioceptive impulses responsible for maintaining balance. Therefore, Parkinson's disease causes reduction of the ductility and as result loss of adaptive reflexes [9]. It should be noted that the population of people with vestibular disorders and the population of patients with Parkinson's disease might be at least partially overlap because pharmacokinetics and pharmacodynamics of antiparkinsonian drugs promotes labyrinth symptoms [10]. Thus, a correlation between Parkinson's disease, disorders of the vestibular system and physical balance shows the importance of the study and balance evaluation of patients with Parkinson's disease.

Objective: development and application of virtual environment for virtual reality devices Oculus Rift DK 2, for assessment of the impact of neurological tests in a virtual reality system to maintain balance of patients with Parkinson's disease.

2 Methods of study

Virtual three-dimensional visual environment is a room decorated in neutral black with bright orange lines and boundaries between floor-wall-ceiling (Fig. 1). This design does not have any distraction, which allows for assessment of the functional state of the system to maintain the equilibrium of the human body.

Figure 1. Three-dimensional environment for assessment of the impact of neurological tests in a virtual reality.

Scenario studies of the vestibular apparatus using virtual reality was designed. To study the possibilities of the proposed method were conducted quantitative estimates of motion vectors when performing routine tests Romberg in the study of neurological status in virtual reality for 20 seconds. The movement data of the body were recorded using markerless motion capture technology. This technique allows recording the three-dimensional coordinates of 20 standard points on the body with a frequency of 15 samples per second. The study involved 12 volunteers without disorders of the balance function; in the second group was 10 subjects with Parkinson's disease. The study of each test included several stages. Stage without glasses: eyes open, eyes closed. Stage with virtual reality glasses: tracking label to the left at 90 degrees for 10 sec., Tracking label to the right at 90 degrees for 10 sec., Tracking label up at 1.75 m. for 10 s., Tracking label down at 1.75 m. for 10 sec., tilt the room forward on 30 degrees , tilt the room back on 30 degrees, tilt the room to the left on 30 degrees, tilt the room to the right on 30 degrees. Evaluation of the path of the points of the body was carried out using the method of integral estimates. The theoretical details of the criterion were published in [13].

At the first step. Motion vector fragment skeleton model as a set of coordinates (t, x, y z) is formed. The whole interval [0, T] then is divided into k fragments with length ΔT. The average values xj, j = 0,. . ., K are calculated inside the [t, t + ΔT] interval. The average values are normalized and stored in a matrix N x k. Where N is a number of subjects in a certain group. Calculation distance-based criterion based on the calculation of the Mahalanobis distance. The theoretical details of the criterion were published in [13]. For the calculation of distance-based criterion Id we have developed a computer program. Input data are files that contain the normalized values of the reference and tested condition.

3 Conclusion

When the Romberg test with open eyes is provided, in both cases there was an impact on the visual analyzer. But in the first case, it was a virtual visual environment while the second - a real room. Compared to the stage of testing with closed eyes the results of these studies showed a smaller deviation integral criterion.

Results of the study of neurological functional tests with closed eyes and using a virtual reality in a group of healthy humans have shown that variations in the point corresponding to the centre of mass were significantly reduced in comparison with the right and left hands. Even so, results of the study of subjects with Parkinson's disease have shown increasing variations in the point corresponding to the centre of mass.

In addition, the characteristics of the obtained values increased integral criterion in the range of 4-14 seconds. The reaction of the receptor cells of the vestibular apparatus, caused by a change in body position in space or a movement leads to a redistribution of reflex muscle tone. Consequently, the time interval from the start of the sample until the maximum value of the integral criterion can be the time of reflex reactions skeletal muscles that maintain the body balance.

References

1. O. S. Levin, N. V. Fedorova, *Parkinson's disease*, 352 (2012)
2. R. Stephenson, A. Siderowf, M. B. Stern, Mov. Disord. **24(2)**, 665 (2009)
3. Y. Miki, F. Mori, K. Wakabayashi, N. Kuroda, S. Orimo, Mov. Disord. **24**, 2299 (2009)
4. C. Gaig, E. Tolosa, Mov. Disord. **24**, 656 (2009)
5. J. Jankovic, J Neurol Neurosurg Psychiatry **79**, 368 (2008)
6. K. J. Brusse, S. Zimdars, K. R. Zalewski, T. M. Steffen, Phys Ther. **85(2)**, 134 (2005)
7. B. N. Smetanin, G. V. Kozhin, Neurophysiology. **39(6)**, 476 (2007)
8. T. Mergner, G. Schweigart, C. Maurer, A. Blumle, Exp. Brain Res. **167(3)**, 535 (2005)
9. V. A. Fokin, Bulletin of the Tomsk Polytechnic University **311(5)**, 120 (2007)

MATEC Web of Conferences 48, 06001 (2016)
DOI: 10.1051/matecconf/20164806001

University educational environment in forming practice – oriented specialist training approach

Mikhail Solovyev[1,a], Irina Abrashkina[1] and Dmitry Kan[1]

[1]National Research Tomsk Polytechnic University, 634050 Tomsk, Russia

Abstract. The article reviews the world experience of forming practice-oriented approach to highly qualified specialist training. Particular attention is paid to the approach of National Research Tomsk Polytechnic University to the organization of practical training of students in the framework of the implementation of World CDIO concept. The authors consider the main results of CDIO concept introduction and implementation of practical training of students, problem situations and their solution, as well as possible ways of further development of the educational environment of the University.

1 Introduction

The enhancing role of technology in economic development and the formation of the political configuration of regions increase the importance of technical specialists in the modern world. In the context of globalization and constant social, cultural, economic and geographic changes it is necessary to pay special attention to the development of personality, the conditions of its formation and self-realization. Higher education is also changing due to the active introduction of information and communication technologies, new teaching formats and methods. The university should create a special environment (the educational environment) which will promote the formation of professional and personal skills and abilities, harmonious education of students, developing their interest in learning, their soft social adaptation, as well as the realization of their desires, abilities and skills acquired.

According to many experts, the creation of such an educational environment is a complex, large-scale and multi-faceted task. The characteristic features of modern educational environment of any technical university are as follows:

- availability of high technology equipment;
- instructors' professional competence;
- continuous improvement of educational content;
- introduction of new teaching methods and educational technologies;
- organization of interaction between educators and trainees, when in joint activities certain relationships contributing to the activity of all participants of the educational process appear.

The following educational environment components are important:

[a] Corresponding author : solo@tpu.ru

- space - availability of modern specialized classrooms, including classrooms with individual attractive design, advanced information and training facilities, modern teaching and laboratory equipment;
- social environment, communication - a system of interaction between students and instructors, high level of communicative culture;
- concept and content of education - possibility of realization of students' individual needs, academic freedom, focus on highly qualified specialist training.

Depending on the university, implemented educational programs, existing scientific schools and other factors, this list can be extended considerably. However, creation of the environment that provides practical training of specialists is an important strategic task for the administrative and teaching staff of all the universities.

2 International experience of the formation of special educational space at technical universities. Theoretical background of practice-oriented specialist training approach

Currently the world educational space shows a considerable number of examples of special educational environment construction, the introduction of non-traditional ways and means of education which are aimed at the development of students especially by applying practice-oriented approach.

General principles of practice-oriented approach to the organization of teaching are:

- interdisciplinary teaching model, focused on students, on the perspective, associated with real technological and social problems, based on real life experience;
- teaching is based on complex projects with active participation of students, which ensures their development and enables them to apply the acquired knowledge and skills;
- most of the information must be obtained by students on their own in the process of studying.

One of most striking examples of practice-oriented approach at technical universities is the Global Initiative CDIO (Conceive - Design - Implement - Operate). The Community of universities with practice-oriented teaching, using CDIO standards, declares and introduces a comprehensive approach to engineering education: a set of common principles for the curricula design, their logistical support, recruitment and teaching staff development.

The CDIO concept originated in the late '90s in the USA in response to the dissatisfaction of employers with the fact that university engineering education is too distant from practice [1]. Officially CDIO community appeared in 2000 thanks to the collaboration of the Massachusetts Institute of Technology (MIT) [2] with three Swedish universities - Chalmers University of Technology, Linköping University and Royal Institute of Technology. The author and co-founder of CDIO Initiative is Edward F. Crawley, professor of aeronautics, astronautics and engineering systems at MIT [3].

The education at universities working according to the CDIO Initiative principles is conducted mainly through practical activity focused on the needs of the external world, rather than on the solution of abstract problems within universities. Graduates from the universities where CDIO standards are applied are highly rated on the international lab our market.

Nowadays more than 100 universities from 25 countries, 10 of them - Russian, such as Scoltech (Moscow), TPU (Tomsk), UrFU (Ekaterinburg), TUSUR (Tomsk), MAMI (Moscow) and others joined the CDIO World initiative. In 2013 the «CDIO RUSSIA» community was established to promote project-oriented approach and the World CDIO initiative standards in educational process.

Each university has its own unique experience of implementing CDIO standards [4], including a unique educational environment formation, based on the construction of an integrated curriculum, special courses and educational modules, organization of students' working space.

2.1 Aalborg University experience (Denmark)

The «Problem Based Learning and Science Technology and Society» course is focused on the development of professional (my profession, language for profession and science, theoretical aspects of profession, professional environment) and social (how to interact with people) aspects. The course includes a number of small projects with mandatory analysis of all the implementation stages and project outcomes. Some examples of projects: "Giraffe" - project on the development of special software for children with autism, project on the conversion of PowerMILL Robot CAM-system for milling programming using an industrial ABB robot.

2.2 Linköping University experience (Sweden)

During the «Engineering project» course (25% of the semester study load) students learn group work, oral and written communication skills. The course implies a compulsory project implementation in a team consisting of 5-6 people. The examples of projects are indoor climate web monitoring system, car operation control system and others.

The project work is carried out under the guidance of the instructor, project results are evaluated by experts. The project is executed in steps: plan, contract that specifies the rules of group work and ways of resolving conflicts, regular reporting to the «customer» according to the design assignment.

2.3 The experience of Moscow State University of Engineering (Russia)

The university is actively implementing STEM (Science, Technology, Engineering, Math) education enabling students to bridge the gap between educational objectives and genuine activities of a scientist and engineer. The examples of conducted STEM-games are «Pandora» (tasks for geology, ecology, biology, botany, geography, astronomy), «Ecological settlement» (targets for transport, pharmaceutical industry, environmental protection, information technology), «Orbit» (tasks for physics, mechanics, electronics) and others.

Compulsory mastering the «Project Activities» course (1st year) involves a variety of engineering projects including those on the employers' orders. The project involves 4 to 12 people receiving education in various majors; the defense of projects is carried out in technical committees.

The best projects are selected by participating in competitions and tournaments like «Engineering start».

3 Experience of practice-oriented educational environment formation at Tomsk Polytechnic University

With the accession to CDIO World initiative Tomsk Polytechnic University (TPU) launched an active and full-scale formation of practice-oriented educational environment in 2011. TPU was the first university in Russia to join this initiative. The University set the task of compulsory introduction of most CDIO concept standards to provide complex engineering training even at a bachelor level.

One of the first decisions at TPU was to introduce CDIO Standard 4 which presupposes the existence of an introductory course which would lay the foundations of engineering practice in product and system development and would be aimed at teaching basic personal and interpersonal competencies [3]. In 2012 TPU developed a new educational module «Introduction to engineering» for all technical educational programs, which is the basis of an integrated curriculum [5].The module concentrates on the acquisition of basic skills and first practical experience that motivate and prepare students for a better understanding of theoretical concepts. Methodically, this introductory course gives the students first impression of engineering features of the major, skills of project work and group communication, which are further developed in the framework of professional disciplines of the program, practice-oriented project implementation and graduate qualification work. The given course is the basis of the practice-oriented training system at the university [6].

Within the module, students are involved in engineering practice by solving problems and simple design tasks, carried out individually and in teams. The introductory course is aimed at stimulating interest and increasing motivation to engineering in students, focusing on practical benefits of basic disciplines.

General structure of a practice-oriented complex engineering training system is presented in table 1.

Table 1. General structure of a practice-oriented complex engineering training system TPU.

Semester	Modules
1	Educational module «Introduction to engineering» (Theoretical part), 1 credit
2-4	Educational module «Introduction to engineering». Creative projects, 3 credits (1 credit per semester)
5-8	Projects within the framework of studying and research work of students
5-8	Complex projects for professional cycle disciplines Preparation of graduate qualification work

The «Introduction to engineering» module consists of two parts: a theoretical part including elements of project activity and a practical part, characterized by a number of team projects carried out under the tutor's guidance. An important feature of the "Introduction to engineering" module is its practical orientation –the formation of students' ideas about their future professional activity and engineering practices in general occurs at the initial stage of training.

The theoretical part focuses on:

- development of students' interest in engineering;
- formation of ideas about future professional activity, engineering practice at early education stages;
- development of basic knowledge and set of skills necessary for solving engineering problems;
- increase in motivation to acquire knowledge and skills in the field of professional training according to the chosen major;
- conscious choice of individual educational path for a particular profile/specialization within a certain major;

The practical part focuses on:

- obtaining deep practical knowledge of professional technical basics;
- development of skills in new product and system creation and operation;
- understanding the importance and strategic value of scientific and technological development of the society;
- acquisition of knowledge about planned educational profile within the major;

The implementation of the theoretical part is based on:

- attracting external specialists in the field of educational programs (leading professors, graduates, employers)
- organization of real (virtual) tours to specialized enterprises, research laboratories, etc.
- solving small theoretical problem-oriented tasks (in the format of an essay, glossary, mini-conferences);
- acquiring project work skills within small project-oriented tasks;

The implementation of the practical part consists in:

- use of team building techniques (team skills, team spirit, team building);
- involvement of external experts in the Main Curriculum implementation (leading professors, graduates, employers) in management;

- solution of short-term practice-oriented problems of applied and research background [7];
- implementation of projects under real production conditions;
- keeping semester journals, drafting project passports, writing analytical reports and analysis on the results of project implementation with advanced IPC support;
- public project defense (with the demonstration of results) with the involvement of external persons concerned;

A sample structure of the theoretical part of the module within "Mechanical Engineering" major is presented in Table 2.

Table 2. Structure of the theoretical part of the module «Introduction to engineering».

№ week	Topics, discussed issues	Type of classes	Forms / teaching methods	Task types / results
1	Introduction to the discipline. General requirements for discipline studying. The initiation of engineering activity, its essence and functions. The development of engineering activity, engineering profession and special education	Lecture	Information lecture	-
2		Practical class	Team skills, team spirit, team building	Group Code
3	The history of training profile 1 ("Information and measuring equipment and technology"), the main directions of educational and research activities, graduates' main customers, practice and employment	Lecture	Overview lecture	Essay on the scientific fields of the training profiles
5	The history of training profile 2 …	Lecture		
7	The history of training profile 3 …	Lecture		
4	Excursion to electrical and electronic engineering laboratories, CAD, mobile object control systems and devices, orientation and navigation devices laboratories	Practical class	Lecture - discussion	
6	Excursion to the thermal control laboratory, radiological material and product testing, acoustic emission methods testing laboratories	Practical class		
8	Excursion to medical engineering laboratory, alternative energy sources (physics of solar elements) laboratory, control devices for cable industry laboratory	Practical class		
9	CONFERENCE WEEK. Theoretical mini-conference	Lecture	Lecture - conference	Group report on scientific subjects of the profiles, group defense in front of the assessment board
10	Basic concepts and definitions in the field of measurement and measurement equipment. Creative projects on profile 1	Lecture	Binary lecture (theorist and practitioner of each profile)	Glossary of measuring devices (distribution of work by letters of the alphabet in groups, compilation of a common glossary)
12	Basic concepts and definitions in the field of non-destructive testing. Creative projects on profile 2	Lecture		
14	Basic concepts and definitions in the field of precision instruments. Creative projects on a profile 3	Lecture		
11	Theoretical and practical principles of project activity. Part 1.	Practical class	Project group work	Map model of an island state (history, flag, population, infrastructure, life

№ week	Topics, discussed issues	Type of classes	Forms / teaching methods	Task types / results
				support, etc.)
13	Theoretical and practical principles of project activity. Part 2.	Practical class		Facility model (house, enterprise, power plant, launch site, etc.). Design, calculations, model work stock. Distribution of roles: designers, foreman, workers, quantity surveyors and others.
15	Debate. The organization of research work of students.	Practical class	Debate	-
16	CONFERENCE WEEK. Practical mini-conference	Lecture	Lecture - conference	A new (own) measurement instrument or device model, preparation of a group report on the basis of semester work, group defense in front of the assessment board.
17	Final practical class	Practical class	Group work	Glossary-based crossword puzzles. Doing rival groups' crossword puzzles
18	Credit test	Practical class	-	Test

The practical part of the module represents the implementation of creative projects for 3 subsequent semesters. The project is an independent solution of a particular teacher-led appliedtask by students or a study on a stated topic. The complexity of the ongoing projects increases from semester to semester and is set in accordance with a list of learning achievements planned for the corresponding semester.

Each semester students are offered a separate list of creative projects with a brief description and a list of the learning outcomes achieved. A list of creative projects is developed by the departments responsible for the profile teaching.

In a real laboratory students learn to solve engineering problems from conception, describing the theoretical (mathematical / physical) model of the future product to its manufacture and operation (they gradually pass through all the stages of engineering activity: «Conceive», «Design», «Implement» and «Operate»).

At the university much attention is paid to the educational environment requirements (CDIO standards 5, 6). Those standards require that [3]:

- students must participate in at least two educational and practical tasks for the design and development of products, one of which would be performed at a primary stage, and the second - at an advanced stage;
- University must have classrooms which would make it possible to organize practical approach to learning creating products, systems and design skills, to transmit disciplinary knowledge, as well as to organize social training.

To meet the requirements of Standards 5, 6 different variants of students' areas-working spaces, intended for project activity organization, have been created at TPU.

4 Results and problems of the formation of practice-oriented educational environment at Tomsk Polytechnic University. Development path

During the formation of practice-oriented educational environment, including the implementation of educational module «Introduction to engineering», a number of problem situations have been faced:

- by the teaching staff (project coordinators):
  - lack of experience of teaching junior students;
  - lack of understanding/awareness of the objectives and processes of a new educational module implementation;
  - reluctance to make changes to the educational process, to his/her own pedagogical activity;
  - unreadiness for innovative activity, especially in the field of applied engineering problem solutions;
  - failure to organize independent work of students;
- at the university:
  - the complexity of scheduling of all the activities related to creative project implementation;
  - low interest/motivation of students;
  - absence of efficient financing schemes of creative projects.

During the educational module implementation from 2012 to 2016 a number of problems were solved. Transparent schemes of all student project activities from the enrollment to graduation have been developed. Several supplementary education programs for teachers, allowing overcoming internal fear of working with junior students, engaging them in real engineering problems, have been designed. Various schemes of student project financing (from departments funds as well as from university funds) have been worked out.

Institutes and departments have started considering the integration of efforts to implement small multi-disciplinary, inter-department educational projects, as well as large-scale interdisciplinary research projects.

5 Conclusion

It should be noted that educational environment of the university with its specificity and structure has a huge impact on training of future professionals. Simultaneously, one of the educational environment quality criteria is its ability to provide all the participants of educational process withopportunities and conditions for effective self-development.Well-organized educational environment is a criterion of success of any university since itforms future generation of specialists, who are directly responsible for their country's future.

References

1. Information from web-site of CDIO. URL: http://cdiorussia.ru (2016)
2. D. A. McAdams, E. T.Morris, ASEE Washington CCS, USA, **113020** (2015)
3. E. F. Crawley, J. Malmqvist, S. Östlund, D. R. Brodeur, and K. Edström, *Rethinking engineering education: the CDIO approach* (Springer, New York, 2014)
4. *CDIO Standards 2.0*. URL: http://cdio.org/implementing-cdio/standards/12-cdio-standards (2016)
5. V. Ivanova, K. Mertins, Mat. Web of Conf. **37,** 01025
6. *CDIO Syllabus2.0*. URL: http://www.cdio.org/benefits-cdio/cdio-syllabus/cdio-syllabus-topical-form (2016)
7. O. M. Zamyatina, P. I. Mozgaleva, M. A. Solovjev, L. À. Bokov, À. F. Pozdeeva, WSJ **27** (ELELC), 434 (2013)

MATEC Web of Conferences **48**, 06002 (2016)
DOI: 10.1051/matecconf/20164806002

Aerospace engineering training: universities experience

Kseniya Mertins[1,a], Veronica Ivanova[1], Natalya Natalinova[1] and Maria Alexandrova[1]

[1] *National Research Tomsk Polytechnic University, 634050 Tomsk, Russia*

Abstract. Contemporary professional working in aerospace engineering must have a set of soft and hard skills. The experience gained in universities shows that training of a competent professional is impossible without an employer involved in this process. The paper provides an analysis of missions, tasks and experience of aerospace professionals and identifies the present and future roles, missions and required skills of a highly qualified specialist in aerospace engineering. This analysis can be used to design a master's program aiming at providing students with the required knowledge, know-how and attitudes needed to succeed as professionals in industrial companies.

1 Introduction

According to the U.S. Department of Labor, Bureau of Labor Statistics, aerospace engineers are expected to have 10 percent growth in employment between 2006 and 2016, about as fast as the average for all occupations.

Aerospace engineers develop new technologies for use in aviation, defense systems, and space exploration, often specializing in areas such as structural design, guidance, navigation and control, instrumentation and communication, or production methods. They often use computer-aided design (CAD) software, robotics, and lasers and advanced electronic optics. They also may specialize in a particular type of aerospace product, such as commercial transports, military fighter jets, helicopters, spacecraft, or missiles and rockets. Aerospace engineers may be experts in aerodynamics, thermodynamics, celestial mechanics, propulsion, acoustics, or guidance and control systems [1].

The aerospace industry is a high-tech industry, which plays an important role in ensuring the national security of the country. The dynamics of the social-economic development of society has led to the necessity of changes in education. Today, the industry needs the specialists who are able to solve complex engineering problems, to meet the future requirements of aerospace projects, to provide the selection of optimal solutions with high technical and economical effect and to take into account long-term forecasts of the impact of technology on the environment.

2 Change in the approach to training specialists

Training for Aerospace Engineering in the late 20th and early 21st century in Russia faced certain stagnation turned into a recession. This was due to the fact that businesses in this area survived rather than evolved since state orders significantly decreased, and the companies were not interested in

[a] Corresponding author : mertinskv@tpu.ru

recruiting young competent professionals. The rise of the industry initiated an interest of employers in competent college graduates; therefore, much attention was paid to the quality of training future engineers. The education standards FSES 3 and FSES 3+ were introduced, according to which the employer and the student became equally involved in the educational activity. Employers became aware of the need to be involved in training of future professionals.

The current state of the aerospace industry generates demand for qualified engineers capable of comprehensive combining various activities, able to solve complex engineering problems, to provide a choice of optimal solutions with high technical and economic effect, to match the foresight requirements of aerospace projects and to take into account long-term forecasts of the impact of technology on the environment and life and activities of both society and man. The formation of engineering competence of the future specialist, task-oriented mastering of graduates in innovative engineering activities, raising the level of social and personal responsibility for the results of professional activities aimed at sustainable development [2] and «green engineering» are becoming of particular importance to ensure the competitiveness of university graduates.

Quality management of engineering education is a strategic objective of engineering education in Russia. It can be provided by the universities (most of which are accredited by ABET) which confirm that the quality of training of technical college graduates for innovation activities depends on many factors: the quality of the educational program; human and scientific potential involved in the educational process; students (including the quality of applicants training); logistical and methodological support; organization of project activities, etc.); educational technologies.

Training of future engineers (including those for space industry) requires practice-oriented training (in accordance with the employer's order) using problem-oriented projects, and it is to be related to real production.

3 Aerospace soft and hard skills requirements

In order to design an educational curriculum in our university, we have to answer two main questions: what knowledge and competences are needed? What does industry expect from such graduates?

It is essential that the professional competence should be integrative and holistic, being the product of vocational training as a whole. As part of the modern paradigm of the education competency, basic methodological approaches to form competences of future engineers are systemic, personality and activity approaches.

Systematic approach allows development of a consistent methodology for formation of the communicative competence based on the activities and functions of the modern engineer. Person-oriented approach provides favorable conditions for self-development of the creative potential of individual and student-centered teaching technologies. Activity approach, in which education and training are based on social experience, focuses on the interaction with the social environment. The efficiency of formation of communicative competence of the future engineer can be enhanced by active learning methods: problem-based lectures and seminars, case-studies, educational and professional simulation games, search techniques, training sessions, group discussions, etc. These methods are specifically encouraging for active mental and practical activities of students thus contributing to the development of their competencies [3]. Table 1 shows the integrative professional competencies of engineers according to Rozhkova's opinion.

Third and fourth year bachelor degree students and first year master degree students, INDT TPU, program "Instrument making" were given an assignment to write an essay "Ideal aerospace engineering training". The students coped better with the description of soft and hard skills the graduates require to possess after graduation than with their proposal of the learning process for formation of these competencies. It should be noted that most of the students wrote that communication competencies and organizational skills are crucial for specialists in this area. Only a quarter of the students wrote that representatives of enterprises should be involved in the educational process. The most often share the idea that lectures should be delivered, on the one hand, by the teacher experienced in the field and, on the other hand, by the teacher who possesses all the necessary modern teaching tools, including advanced training. The majority of students believe that increase in the time practical training in the production will have a positive effect on the quality of training. Some of the students suggested a one-year internship (practice) in the enterprise. According to the students, a good basic knowledge of fundamental disciplines such as mathematics and physics plays an important role. In addition, the respondents note that only highly motivated students who are ready to learn independently and outside the university environment can become competitive specialists. The students expect the university to create an educational environment, including modern laboratories and information environment, and implementation of the best practices of Russian and foreign universities.

Table 1. Professional competences of engineers.

Abilities (skills)	**Content of competences**
Planning and organization	Accuracy, punctuality, control, planning
Leadership skills	Activity, leadership, colleagues respect
The ability to analytical thinking	Analysis, information search
Cooperation	Confidence, sociability
Teamwork	Ability to cooperate, emotional support
Focus on success	Perseverance, following the quality standards
Perception of innovation	Flexibility, initiative
Professional skills	Tutorship, professionalism, self-development
Motivation	Positive attitude to work, energy
Corporate spirit	Loyalty, promotion of corporate values, following the corporative standards
Stress resistance	Attitude towards failures, work under pressure, emotional stability
Independence	Independent opinion, confidence
Oral communication	Oral literate speech, persuasion skills
Writing skills	Informative presentation, literate written speech
Creativity	Ideas generation

4 Experience of Russian and foreign universities

The experience of Russian universities in training of specialists in the field of research development, development of space systems, devices, etc. is implemented in the following forms:

- base department (at the Department of Space Research and Aerophysics MIPT and JSC "Russian Space Systems")
- implementation of educational programs (Department of Space Device Engineering and Innovation Technologies SFU, design and construction of space systems and unmanned spacecraft in the Institute of Radio-Engineering and Telecommunications, Kazan National Research Technical University; ITMO University – for optical and space instrument making, Tomsk State University, specialists in "Design and construction of industrial space systems").

- New educational programs (Moscow State University of Mechanical Engineering (MAMI) in Modern Astronautics);
- fulfilment of master's theses in the JSC "Gazprom Space Systems' (TSU);
- target training of specialists(Russian State Technological University named after Tsiolkovsky (RSTU – MATI) and Moscow Aviation Institute National Research University (MAI);
- scientific internship in enterprises engaged in design, manufacture and operation of space equipment;
- research conducted within the bachelor's degree program, scientific work fulfilled within epymaster's degree program;
- lectures delivered by leading scientists and specialists (including astronauts);
- scientific and practical conferences and workshops;
- participation of employers in meetings of the State Examination Board;
- introduction of the Project Activity Module (MAI, as an example).
- Virtual Design Bureau.

TPU has gained experience in creating a virtual space bureau where students work. The first practice-oriented result was opening of the Mission Control Center. In addition, professionals from the aerospace industry are actively involved in the design of master's and bachelor's degree practice-oriented programs; a joint master's program for the space industry is implemented. Starting from the 3rd year, students take an internship at the enterprises of Roskosmos, and its employees deliver lectures at the university.

In addition, the program of the University development should focus on recruitment and scientific and innovative support of the aircraft, missile and space industries, and other high-tech sectors of the economy.

The analysis of international experience shows that the discussion of the strategy of training experts is to be adopted at state level. As an example, an international aerospace school, where Russian cosmonauts and leading specialists of the Cosmonaut Training Centre named after Yu. A. Gagarin deliver lectures, is organized in Tashkent (Uzbekistan).

The overseas abroad are also actively developing programs to train professionals. Michigan AERO (U-M Aerospace Engineering Department) of the University of Michigan support students' activity for projects in space areas. The training process makes each student, from the freshmen to the graduate, feel part of a strategically important project. The student employed can take a position in the company or lab outside the university, such as NASA, Langley; be involved in the research project or independent study with an Aero faculty member, for credit or pay; take a position as a grader for an Aero class; take a work-study position in Aero [4].

The School of Aeronautics and Astronautics, through Purdue University's Engineering Professional Education program, is offering graduate level courses in aerospace engineering so that students can earn non-thesis MS degrees in Aeronautics and Astronautics entirely via distance learning courses.

Bristol university courses are accredited by the Royal Aeronautical Society. In Southampton, a Space Systems Engineering Course is going to be offered [5].

In addition to Russian forms of training specialists, it is recommended to implement the following measures to enhance practice-oriented engineering training:

- ability to take part in international satellite design competition;
- SEDS (Students for Exploration and Development of Space);
- creation of infrastructure to support student projects (including that intended for development of technology entrepreneurship);
- appointment of AeroFaculty Advisors in the university;
- student chapter of the Institute of Aeronautics and Astronautics;
- development of e-learning (including master degree programs).

According to the survey conducted by N.A. Onishchenko among 440 students of the Aerospace Institute, Orenburg State University, and Kazan National Research Technical University[6], more than

170 students indicate a need for closer integration of theory and practice, 87% of the respondents believe that they will need time to adapt to the workplace irrespective of the high level of training. More than 170 students point out a lack of curriculum interdisciplinary modules and lack of relation to their future profession.

In addition, there is a contradiction between the requirements to the quality of training indicated by employers and insufficient involvement of the enterprises in the design, implementation and assessment of educational programs.

Thus, to improve practice-oriented training, it is necessary:

1. to allocate resources to design innovative educational programs.
2. to provide flexibility, interdisciplinarity and logistics of training modules;
3. to implement network forms for organization of the educational process;
4. to develop strategic cooperation with schools and businesses;
5. to improve the quality of training in fundamental disciplines;
6. to create a career guidance system (starting with primary school);
7. to ensure implementation of space-related research;
8. tointroducepracticalcoursepapersandcreativeprojectsonspacesubjects (offering a wide range of problems and tasks);
9. to solve specific problems of the space industry within the training modules;
10. to introduce modules of technology entrepreneurship;
11. to establish interactive exhibitions and museums with real exhibits, to use prototyping and 3-D modeling equipment in the classroom.

In addition, some changes must be carried out in terms of Human resources: management of educational programs by experts with practical experience, representatives of companies involved in assessment of competencies of the graduates completing these educational programs, transition from the curatorial support to tutors for project activities [7-19].

5 Conclusion

Engineering training in the space area should be the topic of discussion at national level. Only a systematic approach of experts (from classes in children's organizations to the University) with active involvement of enterprises will ensure the results in scientific and engineering solutions of fundamental and applied problems. The activities for development of practice-oriented training proposed as a result of the analysis of the experience of universities (Russian and foreign) will form the environment for regeneration of engineering thinking competencies, soft skills and hard skills in the space sector to train competent professionals.

References

1. Aerospace Engineering Overview. URL: http://www.careercornerstone.org/pdf/aerospace/aeroeng.pdf (2016)
2. N. V. Daneikina, Yu. V. Daneykin, PNV **3-2(43)**, 45 (2015)
3. N.V. Rozhkova, NGU **2 (83)**, 127 (2014)
4. Information from web-site of Michigan AERO. URL:http://www.engin.umich.edu/aero (2016)
5. Space Systems Engineering: course details. URL:http://www.southampton.ac.uk/engineering/cpd/courses/space_course_details.page (2016)
6. N.A. Onishhenko, PhD *Thesis work*, Orenburg, 213 (2007)
7. J. Handelsman, D. Ebert-May, R. Beichner, P. Bruns, A. Chang, R. DeHaan, J. Gentile, S. Lauffer, J. Stewart, J. Tilghman, and W.B. Wood, Science **304**, 521 (2004)
8. D. Cyranoski, N. Gilbert, H. Ledford, A. Nayar, M. Yahia, Nature **472**, 279 (2011)
9. N. Balster, C. Pfund, R. Rediske, J. Branchaw, LSE **9**, 118 (2010)
10. A. Byars-Winston, B. Gutierrez, S. Topp, M. Carnes, LSE **10**, 367 (2011)

11. D. L. Gillian-Daniel, K. A. Walz, Community College Journal of Research and Practice **40**, 145 (2016)
12. S. Hancock, E. Walsh, SRHE **41**, 50 (2016)
13. A. Hajdarpasic, A. Brew, S. Popenici, SRHE **40**, 657 (2015)
14. K. V. Dao, SRHE **40**, 760 (2015).
15. J. A. Douglas, A. Douglas, R. J. McClelland, J. Davies, SRHE **40**, 349 (2015)
16. F. Huang, SRHE **40**, 1492 (2015)
17. S. Brooman, S. Darwent, SRHE **39**, 1540 (2014)
18. N. M. Trautmann, M. E. Krasny, BioScience **56**, 164 (2006).
19. M. Solovyov, I. Abrashkina, D. Kan *Space Engineering 2016* (to be published)

MATEC Web of Conferences **48**, 06003 (2016)
DOI: 10.1051/matecconf/20164806003

Implementation of CDIO Approach in training engineering specialists for the benefit of sustainable development

Yury Daneykin[1,a], Natalia Daneikina[1] and Viktoria Sadchenko[1]

[1]*National Research Tomsk Polytechnic University, 634050 Tomsk, Russia*

Abstract. Due to the competition and changing needs of the world, universities strive to analyze these needs to timely offer necessary educational programs. One of the newest trends in engineering education is sustainable development. To support the concept and implement it into life it is necessary to train specialists able to promote sustainable development that inevitably leads to the necessity to develop educational programs embracing newest approaches considering up-to-date conditions and requirements. There are many efforts globally to develop educational programs for sustainable development. One of the common problems for Universities around the world is the virtual absence of comprehensive educational programs preparing specialists for sustainable development at all three levels of educational process: Bachelor's, Master's and PhD levels. Authors analyze the content of existing educational programs for sustainable development and suggest using *Conceive, Design, Implement and Operate (CDIO) Approach, Standards and Syllabus* that might be successfully applied for the development of up-to-date educational programs for sustainability at all three levels of educational process.

1 Introduction

Modern world has been a global market for quite a long time already. Education is also a product sold around. Potential clients have their own criteria to define the university they would like to study at, but a lot of them study international rankings to be sure in their choice. Project for improving international competitiveness of Russian Institutions of Higher Education with a nominal title "5/100" was initiated in 2013 within the framework of the President's order № 599 "On the measures of realization of state policy in the sphere of education and science". One of the goals is the presence of at least five Russian Universities in top-100 world Universities-leaders to the year 2020 according to International rankings ARWU, QS, THE. There is a number of factors influencing the position of the University in the ranking but for educators the main pride of the University is highly qualified graduates. Two issues of high qualification are under constant analysis first being an educator's concern how to train a good specialist and the second is what specialist are on demand for the global development.

[a] Corresponding author : daneykin@tpu.ru

2 Authors artwork

One of the newest trends in engineering education is the concept of sustainable development. The idea of sustainable development originates from numerous environmental movements in earlier decades and was defined in 1987 by the World Commission on Environment and Development (Brundtland Commission 1987) as: "Development that meets the needs of the present without compromising the ability of future generations to meet their own needs" [1]. To support the concept and implement it into life it is necessary to train specialists able to promote sustainable development that inevitably leads to the necessity to develop educational programs embracing newest approaches considering up-to-date conditions and requirements. The term "Education for sustainable development is used by the United Nations Organization and all international community. As a recognition of the importance of education for sustainable development General assembly of UNO declared the years 2005-2014 as a decade of education for sustainable development with the aim to transfer to sustainable development through all forms of education, social consciousness and highlighting the important role of education and learning. Education for sustainable development implies a synergy of disciplines targeted at research and solution of global problems in resource efficiency, satisfaction of human needs while keeping the environment so that these needs will be satisfied not only for the present but also for future generations. There have been quite a lot of work done by educators around the world to analyze the insight of engineering educational programs for sustainable development and find approaches for training engineers ready to promote the concept of sustainable development. Some colleagues find it necessary to cooperate tightly with industries to be perfectly aware of their need and structure educational programs accordingly [2]. American researches found their way adapting 12 principles of Green Engineering that can be the basis for educational programs for sustainable development [3] and analyzing how comfortably they are incorporated into the education for sustainability. They also indicate the problem common for Universities around the world – the virtual absence of comprehensive educational programs preparing specialists for sustainable development at all three levels of educational process: Bachelor's, Master's and PhD levels.

In addition, the following elements can be considered as essential when developing educational programs.

- Opportunity for fundamental research to receive new knowledge on the process of transition to the sustainable development of modern society.
- Opportunity for applied research for the creation and promotion of the culture for sustainable development, global system thinking in the interests of sustainable development.
- Development of regional programs for sustainable development taking into account local peculiarities of environment, social and economic factors, cultural aspects, industry, population etc.
- Detection of the systems of indicators of sustainable development, analysis of the data and opportunity to track the progress on the path to sustainable development.
- Problems and prospects of transition to sustainable development.

We analyzed what educational programs exist on the example of leading Universities according to ranking Times Higher Education-QS 2013–2014 [4] (Table 1).

Table 1. Educational programs on the example of leading Universities.

Washington University in St. Louis	The University positions itself as a national leader in the sphere of sustainable development. It offers Bachelor's educational program providing interdisciplinary approach to understanding complex ecological, economic and social problems. Students look inside sustainable life style from different points of view including scientific, political, economic, social and philosophic. Educational program embraces three main directions in sustainable development: ecological, business strategies for sustainable development and design of sustainable spaces. Program highlights development from local to global and considers the quality of life of a person, efficiency of institutions and safety of the planet.

The University of Hong Kong	The University provides several courses in sustainable development. One of them encompasses general issues of sustainable development; the second is devoted to "smart" architecture. Students of the third course explore projects of the future and research interactions between environment and social contexts. One more course is mainly concerned with energy issues with some aspects of climate change.
Peking University	The University has an experience of conducting summer seminar "Global sustainable development". The seminar involves discussing sustainable development in technical, social, economic and ethical aspects. Various issues are under discussion, such as climate change, human health, clean water, energy, biodiversity, ecosystems, poverty, human resources, safety, globalization and economy.
Boston University	The University offers certificate program in "Sustainable development in energy" which includes courses in business studies, economy, politics and engineering disciplines.
Princeton University	The Program in Sustainable Energy is designed for Princeton undergraduate students who are interested in pursuing careers or graduate education in the area of Sustainable Energy Science and Technology. The program includes courses in energy resources, technologies and their influence on climate and environment, development of innovative energy systems, understanding of global climate. In addition, it puts accent on joint research in the sphere of renewable energy sources and alternative fuels.
Cambridge University	The Prince of Wales's Business & Sustainability Programme has been developing for more than twenty years and serves a strategic forum for top leaders for the study of innovative and pragmatic approaches for sustainable development.
Imperial College of London	Master's program in "Sustainable development Futures" is aimed at the development of new generation of leaders in energy sector and designed mostly for the students with knowledge of Physics. The course gives fundamental knowledge in global issues of energy industry and sustainable energy technologies.
University of California	The University offers the program "Leaders in Sustainable Development". Economic, ecological and social factors are considered as key elements of sustainable development. One of the advantages of the program is facilitating students' cooperation with different specialists that can help looking more widely at the problems of sustainable development.
Johns Hopkins University, Whiting School of Engineering	Minor educational program "Engineering for Sustainable development" is concerned with poverty, social inequality. The specialist graduating this program will cooperate with a wide circle of organizations and subjects including national and local authorities, non-governmental organizations and local communities. Engineering program students learn approaches to data gathering in various conditions.
University of Wisconsin–Madison	Master's educational program "Sustainable development of Engineering Systems" prepares for work with complex systems and their influence on the quality of water, earth, air, energy, economy and society.

We can see from the table that mostly educational programs for sustainable development focus on social, energy and environmental factors. Some programs imply global approach to the issues of sustainability, while others can concentrate on local needs, like smart architecture for Hong Kong. Even leading Universities do not offer educational programs in all three levels: Bachelor's, Master's and Postgraduate's degrees. So it necessary to find approach that can be used for developing complex engineering educational programs.

Many universities around the world use the CDIO approach (Conceive, Design, Implement, Operate) to update their Bachelor's (BEng) programs to prepare graduates to more complex engineering activities at all stages of the life cycle of technical objects, processes and systems [5]. This approach is widely used, as it is consistent with the requirements of international standards (IAE Graduate Attributes and Professional Competences) to the engineering HEI's graduate learning outcomes and competences of professional engineers (www.ieagreements.org/IEA-Grad-Attr-Prof-Competencies.pdf). The CDIO approach allows the design and implementation of BEng programs as a basic engineering education in accordance with the criteria for accreditation of engineering programs in the countries - signatories of the Washington Accord, including the accreditation criteria of the Association for Engineering Education of Russia (AEER) [6]. In some Universities "CDIO concept was regarded as a new model for engineering education whose aims were to develop students' ability of engineering application to solve their employment problems efficiently" [7].

Today, more than 100 universities from 25 countries joined the World initiative CDIO, 10 of them - Russian, such as Skoltech (Moscow), TPU (Tomsk), URFA (Ekaterinburg), TUSUR (Tomsk) MAMI (Moscow) and many others.

Some features of these universities:

• 9 of them are simultaneously present in the QS rankings, THE, ARWU, (including 5 TOP-100) 45 simultaneously in 2 rankings of 3;

• 44 are in the TOP-500 QS rankings, TPU takes 485 place;

• 33 are in the TOP-300 rating in THE, TPU takes 251-300 place;

• 10 are in the TOP-100 rating in ARWU.

In 2013, «CDIO RUSSIA» academy was founded to promote the project-oriented approach to education and facilitate educational process in accordance with the standards of the World CDIO Initiative.

In October 2011, TPU joined the worldwide CDIO concept, becoming the first Russian university that adopted the initiative of leading universities in the modernization of engineering education. A series of workshops were held to explain the basic principles of the concept of modernization of engineering education, including those at international level (The Royal Institute of Technology, November 2011).

There are 12 Standards offered by CDIO Approach [8]:

STANDARD 1: The Context
STANDARD 2: Learning Outcomes
STANDARD 3: Integrated Curriculum
STANDARD 4: Introduction to Engineering
STANDARD 5: Design-Implement Experiences
STANDARD 6: Engineering Workspaces
STANDARD 7: Integrated Learning Experiences
STANDARD 8: Active Learning
STANDARD 9: Enhancement of Faculty Competence
STANDARD 10: Enhancement of Faculty Teaching Competence
STANDARD 11: Learning Assessment
STANDARD 12: Program Evaluation

These Standards indicate a clear direction for the development. In December 2011, TPU authorities approved the list of "pilot" educational programs for modernization in accordance with the CDIO standards: Technological machinery and equipment, Chemical Technology and Power and Electrical Engineering. All actions for updating "pilot" programs were grouped into three main areas.

Comprehensive modernization of educational programs begins with the definition of the status based on evaluation of compliance of general educational programs (GEP) with CDIO Standards. Concept, objectives, learning outcomes, content, and technology of implementation of GEP should be adjusted according to the nature and content of future engineering activity of the graduates. Developers of educational programs expect various learning outcomes including fundamental knowledge and acquisition of practical skills, formation of personal and interpersonal qualities of graduates and developing their responsibility for economic, social and environmental impact of engineering activity.

Organization of workspace implies a gradual and systematic improvement of logistical and laboratory facilities of the University, as well as the creation of "workspace islands" for practice-oriented training, for team and independent activity of students.

The development of competencies of teachers involves improving their skills in the field of engineering activities (training at the enterprise sector) and in the area of problem-oriented and project-organized learning technologies (seminars and training in the CDIO-participating universities).

Designing educational programs using the developed Standards can guarantee comprehension and high quality of the programs as they embrace different aspects and look deep inside the content, pedagogical, and methodological sense. Standards cover content aspects and results, requirements to qualification of both students and professors, touch upon the quality of working environment, and involve teaching methods.

The content-related part of educational programs can be designed using CDIO Syllabus. If we analyze the Syllabus [9], we can see how it can be adopted to the programs on Sustainability.

The first part of the Syllabus is devoted to Disciplinary knowledge and reasoning, which includes knowledge of underlying Mathematics and Sciences (with Mathematics, statistics, Chemistry and Biology inside), core engineering fundamental knowledge and advanced engineering fundamental knowledge, methods and tools. As we can see this part is highly compatible to the idea of education for sustainability. The specialist in sustainability, especially looking at environmental, climate and energy aspects is supposed to acquire definite scientific background.

The second part of the Syllabus embraces personal and professional skills and attributes. Uncovering the inside we can see the following items: analytic reasoning and problem solving, experimentation, investigation and knowledge discovery, system thinking, attitude thoughts and learning, ethics, equity and other responsibilities. All these aspects are on high demand for the specialist in sustainability as providing the specialist with skills both useful and necessary. It might be topical to decide to what degree these aspects should be developed at all three levels of education [10, 11]. For example developing responsibility is crucial at all three levels as each specialist ought to be responsible for the part of duties they undertake. As for analytical reasoning and investigation, they might be on more advanced levels for Master's and Postgraduate's degree students.

Third part of the Syllabus includes Interpersonal skills: teamwork and communication. Some of the programs of leading Universities listed above stress the necessity of cooperation of a specialist in sustainability with other specialists and developing teamwork and communication skills will prepare students to more fruitful collaboration in the future with the colleagues and organizations.

The forth part includes conceiving, designing, implementing and operating systems in the enterprise, societal and environmental context – the innovation process (with external, societal and environmental context, enterprise and business context, conceiving, system engineering and management, designing, implementing and operating inside). The compliance of this part of the Syllabus with the skills and knowledge of a specialist in sustainable development is obvious. Although, it might be disputable in what volume these knowledge should be given to specialist at all three levels of education. It seems reasonable that conceiving is more a prerogative of postgraduate student while operating will be more a function of a Bachelor.

3 Conclusion

The concept of sustainable development is one of the newest trends in engineering education. To support the concept and implement it into life it is necessary to train specialists able to promote sustainable development that inevitably leads to the necessity to develop educational programs embracing newest approaches considering up-to-date conditions and requirements.

Education for sustainable development implies a synergy of disciplines targeted at research and solution of global problems in resource efficiency, satisfaction of human needs while keeping the environment so that these needs will be satisfied not only for the present but also for future generations.

One of the common problems for Universities around the world is the virtual absence of comprehensive educational programs preparing specialists for sustainable development at all three levels of educational process: Bachelor's, Master's and PhD levels.

Existing programs can be clearly divided into the following types: Bachelor's programs, Master's programs, certificate courses, summer seminars. There are programs that cover quite many aspects of sustainable development including technical, scientific, economic and political issue, while others have a narrower sphere (for example "smart" architecture). Nevertheless, it is still a necessity to develop a comprehensive approach to the design of educational programs at all three levels of education: Bachelor's, Master's and Postgraduate degrees.

Many universities around the world use the CDIO approach (Conceive, Design, Implement, Operate) to update their Bachelor's (BEng) programs to prepare graduates to more complex engineering activities at all stages of the life cycle of technical objects, processes and systems and this Approach, including CDIO Syllabus, might be successfully applied to develop up-to-date educational programs for sustainability at all three levels of educational process with some consideration of the extent to what different knowledge and skills should be given and developed at each stage of educational process. CDIO Standards may serve as a guarantee for the high quality of educational programs.

References

1. Official site of Global Issues Organization. URL: http://www.globalissues.org/article/408/sustainable-development-introduction (2016)
2. G. Kruss, et al. Int. J of Edu. Dev. **43**, 22 (2015)
3. C. Murphy, et al. Environ. Sci. Technol. **43**, 5558 (2009)
4. Yu. Daneykin, N. Daneikina, Priv. Sci. Bul. **41**, 41 (2015)
5. A. Chuchalin, *Proceedings 8th International CDIO Conference* (Queensland University of Technology, Brisbane, 2012)
6. A. Chuchalin, M. Tayurskaya, J. Malmqvist, Eur. J. of Eng. Edu. (to be published)
7. J. Bai, et al. *Procedia - Social and Behavioral Sciences* **93**, 381 (2013)
8. CDIO organization. URL: http://www.cdio.org/implementing-cdio/standards/12-cdio-standards (2016)
9. CDIO organization. URL: http://www.cdio.org/benefits-cdio/cdio-syllabus/cdio-syllabus-topical-form (2016)
10. A. Chuchalin, N. Daneikina, *CDIO Proceedings* (to be published)
11. V. Ivanova, K. Mertins, Mat. Web of Conf. **37,** 01025

MATEC Web of Conferences 48, 06004 (2016)
DOI: 10.1051/matecconf/20164806004

The status of world outlook and environmental competencies in aerospace engineering training: the general and special

Igor Ardashkin[1,a], Alexander Bykov[2] and Ekaterina Kirsanova[1]

[1] National Research Tomsk Polytechnic University, 634050 Tomsk, Russia
[2] National Research Tomsk State University, 634050 Tomsk, Russia

Abstract. The article is devoted to the study of the status of philosophical and environmental competences for engineering education programs in the field of space exploration. The authors prove that the specialists who will work in the space field in the XXI century face the most important problems, the solution of which is vital for humanity. It is stated that the quality of future decisions will be determined by the development of the philosophical and environmental ideas. It is demonstrated that these idea should necessarily have futurological character. It is suggested that the special courses aimed at the formation of futurological philosophical and environmental concepts can present as a possible solution to this problem (e.g. Creativity training for engineers, etc.).

1 Introduction

Nowadays space is one of the most important areas for scientific knowledge and the development of human civilization. The future of humanity is largely determined by the level of prospective success of space development. Space has enormous latent possibilities, penetration into of which is not just a cognitive interest but vital task for human civilization that determines the fate of it. Especially, today when humanity is on the threshold of space interaction new phase (industrial exploration) [1]. It is connected with the development of consecutive penetration into space, and with the research of evolutionary processes of global scale. Obviously, the processes that occur on our planet and lead to the emergence of serious global problems (global warming, nutrition problems, poverty, ecological disruption etc.) require the appropriate supervision and management, the transformation of these processes from a class of dangerous for life and social activity to a class of neutral or advancing the conditions of human existence and society.

To solve the above mentioned problems, it is important to develop projects of space exploration, estimating space not only as the scope of scientific research, but also as the sphere of economic and social potentials. Due to this, it is necessary to establish the subject oriented educational programs in higher educational institutions. This does not mean that such specialists are not trained nowadays in various universities around the world, but it is expected that the quality of such educational programs meets the needs of latest challenges of the time, the essence of which was mentioned above [2].

Russia has wide experience of personnel training for space industry. This experience can be represented as one of the leading in the world, taking into account the achievements in the field of

[a] Corresponding author : ibardashkin@tpu.ru

space exploration in our country. But today, there are several problems in this area connected with the lack of specialists for enterprises providing the activity of space industry. Usually these problems are related to an age gap of specialists (when very young professionals aged from 22-23 years and very senior professionals over the age of 60 years work together), to loss of work attractiveness in the space industry for economic and social reasons, to the degree of image and to the reform of educational standards etc.

With a glance to the current situation it is necessary to appeal to the quality of personnel training for the space industry on the example of Russian Federation. Moreover, the authors try to focus primarily on determination of role and importance of philosophical and environmental competences, believing that the development of these abilities of future experts will determine the quality of their training in general [3, 4]. This interconnection is, primarily, reasoned by the specifics of cosmic sphere – the space that is the most different from the life conditions of our planet, and therefore it represents for people (humanity), the habitat, features and functions of which absolutely other than our usual terrestrial conditions of existence. Hence, the experts who are preparing to explore, to investigate and to be in outer space should take into account as much as possible its specific features in their professional activities, and the task of education, in particular, is to form such competences [5].

2 Results and discussion

Appeal to the issue of the status assessment of philosophical and environmental competences of space exploration of engineering training is determined by several aspects. These aspects are of the different nature: philosophical, cognitive, socio-economic, legislative, environmental, technological, etc.

The main aspect of updating of this topic is current situation on our planet, which is fatal for humanity. The solution of existing and actively growing problems is impossible without active exploration and development of outer space [6, 7]. With it can be connected the identification of approaches to address critical issues of our time. It must be obvious that research in this field has a radically different character and origin, than conventional research on the Earth. For these reasons, research requires a completely different approach due to the fact that a researcher is in a situation of critical uncertainty and cannot always clarify what is happening, especially when describing it in terms of the existing research. Often, some phenomena can be neglected by scientists due to the fact that they do not consider them noteworthy. The situation can often resemble the reality of a fairy tale, where the researcher is required to «go somewhere and to bring something». This reason is more of philosophical nature and is the most general in comparison with other reasons. That, however, does not set aside the necessity to indicate them.

Another aspect of the appeal to the status assessment of philosophical and environmental competences of engineering training of space exploration stems from the first one, but it has independent significance. This is an aspect of cognitive character. The cognitive aspect distinguishes the specificity of human cognitive practice, draws attention to the organization of cognitive process. The fact is that a person within the boundaries of existing terrestrial space, in the process of scientific knowledge usually had to deal with the study objects, available to his perceptive apparatus. In the cognitive sciences, this type of object was named human-dimensional. But when people started to investigate, the microworld (quantum reality) on the one hand, and, they discovered megaworld (space and the universe) on the other hand, then the question about the perception of these worlds arose, because these objects of cognition became non dimensional for human. In order to cognize, to imagine this type of reality a person had to fall back upon to all the efforts, going beyond his empirical possibilities. This reality has been transformed into a mathematical formula, or fantastic images, unrelated to understandable and accessible meanings. This reality was of paradoxical nature. Moreover, in the cognitive sense, the problem of the Unity of the world started to manifest, which was previously testified by scientific principles obtained in physics, (in particular, Heisenberg's uncertainty principle, Bohr's principle of complementarity, etc.).

Many researchers even started to speak about the formation of a new type of thinking, which began to reflect all the above mentioned transformations. It is proposed to call this type cosmic. Particularly, according to Russian researcher L.V. Shaposhnikova, "new cosmic thinking requires a different system of cognition and other methodological foundations different from those that exist in modern theories of knowledge. Understanding and establishing the new system, there is an urgent need for further development of cosmic thinking ... Space in the context of new thinking is considered, not only as an astronomical concept, but in all its energy wealth and diversity of states of matter ... This system of knowledge will be closely connected with the moral and ethical issues ... Cosmic thinking includes a new approach to the study of human problems as a cosmic structure in which the phenomenon of consciousness is an important evolutionary orientation" [8, 9].

We may not agree with the name of a new type of thinking, we may supplement its existing features with others, but we cannot deny the need of these changes. And the fact that the things which were identified there as nontraditional features of cognitive processes has a frivolous nature, because in this article it is important to indicate this trend without its detailed analysis.

Accordingly, these cognitive transformations cannot be ignored nowadays while engineers training related to the space industry.

A third aspect of status assessment of philosophical and environmental competences of engineering training in the field of space is the fact that today humanity has launched a new stage - the stage of industrial development of space associated with the need to use space resources, outer space, energy, etc. for the benefit of humanity. This assessment is also confirmed in literature.

Particularly, S.V. Krichevsky points out that "at the present time in the field of space activity the transition to the industrial stage is carried out. This stage is represented by the creation of global space systems and technical infrastructure in space, by the solution of vital for humanity actual and future practical tasks of extraterrestrial resources exploration. In this regard, Russia and world community should develop and implement adequate policy (cosmic, scientific, technical, economic, environmental), aimed at natural resources management and natural environment protection in order to ensure sustainable and safe development. This requires the establishment and implementation of appropriate "laws of the game" and administration systems, covering the totality of corresponded relations [7].

Such branch of development should be related to the two previously presented aspects as long as industrial exploration of space must have a fundamentally different nature than its realization on the Earth. Space should become an integrating factor for humanity, should contribute to understanding of the unity of mankind in social and economic development of space. Without this condition it is very difficult to imagine the future of humanity, because if those existing nowadays issues of political, economic, social, religious and ethnic nature remain in the process of space industrialization, it will have a negative impact on the fate of our civilization in general. This integration is needed not only because of ethical and moral reasons associated with a sense of responsibility for their own destiny, but also because of the fact that no country is able to develop and maintain the technological component of industrial space exploration individually. All countries, societies and cultures should contribute to this process, without mutual efforts the process will be impossible. The experience of the international space station clearly demonstrates it [6].

Legal aspect plays important role in the issue of status updating of philosophical and environmental competences of engineering training in space exploration. The fact is that today, the space serves as a place, which is accessible to absolutely every one of our planet, because space expanses are not legally fixed. At the present moment it facilitates the process of space exploration, but if there is an incoordination between the countries about the conditions and rules of exploration and work in space, this legal void can play a negative role. Therefore, aspects of the legal regulation of space exploration are essentially significant and this should be reflected in consciousness of experts, whose professional activity is connected with space. In addition, the integration of human communities on the Earth facing the necessity of space exploration will be more solid if it is carried out on a legal basis, if it governed by legal mechanisms.

The ecological aspect also plays an important role in status assessment of philosophical and environmental competences of engineering training in space exploration. Also it is significant because in relation to space environmental issues play a dual-use role. Firstly, these aspects are extremely important for the planet, which is experiencing the effects of the most powerful pollution of its history and ecological disruption. Secondly, penetrating into space, humanity risks repeating the negative experience in the new space, which seems, though infinite, but unpolluted. It does not dispense with the necessity of awareness of the problem in time (especially that even nowadays near-Earth space is contaminated enough by so-called "space junk").

According to S.V. Krichevsky, "regulation of extra-terrestrial natural environment is clearly lagging behind space activity that is growing on a set of directions and dimensions, including those already outside the near-Earth space environment (for example, in connection with landings to Mars and plans of its colonization). This reflects not only formal legal relationships, but also an outlook and psychology of people on space activities in extraterrestrial natural environment" [7].

The repercussions of these aspects affect the system of engineering education that is pointed out by many foreign experts. In particular, J. Halb, J. Adamovsky, K. Paul-Vosty state, «this nature of the engineering method is also reflected in the evolution of engineering curricula. In recent years, traditional engineering education in such areas as material science and construction has been complemented with education in ecology, economics, stakeholder participation or ethics in order to react to new challenges in the engineering profession.

One of the more recent challenges in engineering practice and education is the consideration of sustainable development issues (as well as related issues such as climate change, increasing complexity, etc.). Sustainable engineering comprises a life-cycle perspective and consideration of ecological, economic, and socio-cultural aspects [10].

It is obvious that remaining technological aspect of philosophical and environmental competences of engineering training in space exploration should converge all other presented aspects in space engineering and technology development [11]. It is essential that trained expert in the field of space exploration really gets through the mastery of philosophical and environmental competence the image of the area where he has to work. These aspects related to the ideological and environmental competences of educational programs may be described as futuristic as they all involve a certain "flash-forward" into the future, the ability to see something, which does not exist yet, that, how equipment and technology of new type can operate in uncertain conditions.

It is analyzed needs that make us apply to educational standards relating to the preparation of specialists for cosmic sphere in order to find out if they have competences required. To be objective, let us consider the standards of different levels of Bachelor's program (161100 "Guidance, navigation and control"), Specialist's program (160400 "Design, production and operation of rockets and space-rocket complexes"), Master's program (160400 "Missile Systems and Space") [12-14]. The educational standard of bachelor educational program includes 16 general cultural competences, specialist's program - 17 general cultural competences, the standard of master's program - 17 general cultural competences. If you compare the content of these competences with the content of similar competences, but of the standards of other engineering programs not related to space exploration, you can find out that contents are almost identical (e.g. 140400 Power and Electrical Engineering (Bachelor) 130101 Applied Geology (specialty), 04.15.01 Mechanical Engineering (Master)) [15-17]. This means that the futurist aspect of philosophical and environmental competences of the preparation of specialists in the field of space exploration is not taken into account [18].

According to the authors of this article, this is an important omission because the fate of mankind and the earth civilization depend on the activity of the future specialists to a large extent. In this article the authors are not ready to offer a full set of competences of futurological plan, but they are ready to name several options for a possible solution of this problem. In particular, in 2016 within the educational bachelor program of Tomsk Polytechnic University the course "Creativity training for engineers" was introduced. One of the most important tasks of it is the development of creative competences, imagination of students within the framework of above mentioned futurological format. If students do not generate images of reality beyond the ordinary boundaries, if they are not forced to look for unorthodox solutions, then during their professional activities they will not be able to find necessary engineering solutions, especially when it comes to the scope of space. It is possible to offer a similar course aimed at developing the environmental competencies, which will focus on the development of ecological concepts and skills of extraterrestrial type (for example, extraterrestrial space ecology).

To confirm this idea, we would like to appeal to historical analogy related to the development of space industry in our country. Russia (USSR) launched the first artificial satellite into space, the first man, they first came out in outer space, etc. not only because enormous financial, material and human resources were used, but also because, long before these events in our country new direction called Russian cosmism was widely practiced. For contemporaries the ideas of N.F. Fedorov, K.E. Tsiolkovsky, V.I. Vernadsky and others seemed especially inexplicable and inapplicable, fantastic, but with the help of them people were forced to think about the connection of the Earth and space, about the possibility of penetrating into space. As noted by L. Margulis, D. Sagan, "it is impressive how Vernadsky erased rigid distinction between living organisms and inanimate environment, describing life globally, even before the first images of the Earth surface from space. In fact, Vernadsky created for the space the same thing as Darwin for the time: as Darwin showed that life arose from a common ancestor, so Vernadsky showed that life is "built-in" in unbreakable indissoluble material - biosphere. Life is a community processing energy from the sun into earth substance" [19].

3 Conclusion

To conclude with, it should be noted that the status of philosophical and environmental competences of engineering training of specialists in the field of space exploration will play one of the defining roles. This conclusion is due to the nature of the professional activity of such specialists, according to which they will prepare the technological developments and solutions to an environment that is radically different from the Earth. Therefore, futurist aspect of philosophical and environmental competences of future experts will have to determine the nature of their futuristic technological solutions and developments. As one of the possible ways in order to form such competences the authors propose to organize special courses aimed at creating futurological context of educational preparation (e.g. Creativity training for engineers).

Acknowledgement

The article is supported by the Russian Humanitarian Science Foundation grant.

References

1. B. S. Caldwell, Act. Astr. **112** (2015)
2. J. Rodriguez, A. Laveron-Simavilla, J.-M. del Cura, J. M. Ezquerro, V. Lapuerta, M. Cordero-Gracia, Adv. In Sp. Res. **56**, 7 (2015)
3. S.P. Kurdyumov, *Mater. mezhd. nauch.-obshch. konf.* **1** http://lib.icr.su/node/877
4. G.D. Dulnev, *Mater. mezhd. nauch.-obshch. konf.* **1** http://lib.icr.su/node/885
5. B. Donald, Act. Astr. **107** (2015)
6. P. Dias, Scien. and Engin. Ethic. **20**, 2 (2014)
7. T. Thumm, J.A. Robinson, Act. Astr. **103** (2014)
8. S.V. Krichevskiy, *Osvoenie vnezemnykh prirodnykh resursov I okhrana okruzhayushchey sredy. Upravlenie sotsioprirodnymi sistemami: filosofsko-metodologicheskie aspekty* (Rossiyskiy arkhiv gosudarstvennykh standartov, Moscow, 2006)
9. L.V. Shaposhnikova, *Mater. mezhd. nauch.-obshch. konf.* **1** http://lib.icr.su/node/876
10. I.B. Ardashkin, L.A. Korobeynikova and A.V. Popova, Mat. Web of Conf. **37** (2015)
11. J. Halbe, J. Adamowski, C. Pahl-Wostl, Jour. of Clean. Prod. **106** (2015)
12. R. Correal, G. Pajares, J.J. Ruz. Robot. **34**, 2 (2016)
13. *Prikaz ob utverzhdenii federalnogo gosudarstvennogo obrazovatelnogo standarta vysshego obrazovaniya po napravleniyu podgotovki 161100 Sistemy upravleniya dvizheniem I navigatsiya.*

http://www.consultant.ru/document/cons_doc_LAW_192521/2ff7a8c72de3994f30496a0ccbb1dd afdaddf518/ (2016)
14. *Prikaz ob utverzhdenii federalnogo gosudarstvennogo obrazovatelnogo standarta vysshego obrazovaniya po napravleniyu podgotovki 160400 Proektirovanie, proizvodstvo i ekhspluatatsiya raket I raketno-kosmicheskikh kompleksov.* http://www.edu.ru/db-mon/mo/Data/d_11/m82.pdf (2016)
15. *Prikaz ob utverzhdenii federalnogo gosudarstvennogo obrazovatelnogo standarta vysshego obrazovaniya po napravleniyu podgotovki 160400 Raketnye kompleksy I kosmonavtika.* http://www.consultant.ru/document/cons_doc_LAW_98340/5d5349cf7e1bf3869c85e8494a0df91 25b321608/ (2016)
16. *Prikaz ob utverzhdenii federalnogo gosudarstvennogo obrazovatelnogo standarta vysshego obrazovaniya po napravleniyu podgotovki 140400 Elektroenergetika I elektrotekhnika.* http://www.edu.ru/db-mon/mo/Data/d_09/prm710-1.pdf (2016)
17. *Prikaz ob utverzhdenii federalnogo gosudarstvennogo obrazovatelnogo standarta vysshego obrazovaniya po napravleniyu podgotovki 130101 Prikladnaya geologiya.* http://www.edu.ru/db-mon/mo/data/d_11/prm62-1.pdf (2016)
18. *Prikaz ob utverzhdenii federalnogo gosudarstvennogo obrazovatelnogo standarta vysshego obrazovaniya po napravleniyu podgotovki 15.04.01 Mashinostroenie.* http://cdnimg.rg.ru/pril/107/88/52/35179.pdf (2016)
19. I.B. Ardashkin, A.N. Yakovlev, N.V. Martyushev, Adv. Mat. Res. **1040** (2014)
20. L. Margulis, D. Sagan, *Chto takoe zhizn? V.I. Vernadskiy: Pro et Contra* (Russkiy christianskiy gumanitarnyy institut, St-Petersburg, 2000)

MATEC Web of Conferences 48, 06005 (2016)
DOI: 10.1051/matecconf/20164806005

Active learning approach in Moodle for the organization of student's self-study practice-based learning activities

Veronica Ivanova [1,a], Kseniya Mertins[1], Mariya Abdrashitova[1], and Dariya Isaeva[1]

[1] National Research Tomsk Polytechnic University, 634050 Tomsk, Russia

Abstract. Nowadays e-learning tools and delivery methods have been constantly expanding. Employs use e-learning to train their employees more often and often. New and experienced employees have the opportunity to improve upon their knowledge base and expand their skill sets. At home, individuals are granted the access to the programs that provided them with the ability to earn online degrees and enrich their lives through the expanded knowledge. The paper focuses on the analysis of the advantages and disadvantages of e- learning. The ways of applying on-line training used by employers are demonstrated. The experience of implementing active methods of e-learning is described as well as the conclusion about the possibility of their application is made. The paper also presents the results of the survey conducted among TPU teacher and students concerning the advisability of e-learning usage.

1 Introduction

The current stage of the development of active training methods is characterized by the increasing distribution of interactive training technologies. Distinctive features of active and interactive training are reflected in the way of organizing the interaction of the teacher and students. The main way of interaction becomes not only active feedback between the teacher and students, but also interaction of students among themselves. The main functions of the teacher have changed — the teacher acts more as an organizer and an assistant, collaborative learning in small groups becomes the main form of training.

Modern telecommunications technology allows students and teachers to support a constant "live" (interactive) dialogue with a real partner, and makes it possible to communicate in real-time.

E-learning environment has changed the delivery methods of training materials as well as the character of interaction among the participants of the educational process. In different models of e-learning the share of tasks performed in the conditions of direct contact "face to face" between a teacher and a student decreases. This fact influences the choice of teaching methods.

The educational environment, in which e-learning is carried out, is interactive by its nature. It is the Internet with its great opportunities for creation and transfer of educational content, the organization of interaction and cooperation. It allows teachers to find and choose the most appropriate tools and technologies by means of which it is possible to apply methods of active training in teacher-led training sessions and in the on-line environment.

[a] Corresponding author : kurator@tpu.ru

Communication activities (e-mail, forums, conferences, on-line blogs, etc.) and exploration activities (mainly navigation of contents) are required by e-learning. These activities can be found in a Learning Management System (LMS). LMS is a platform for administrating, documenting and delivering e-learning contents, which offers the enrolled students a vast number of courses with high customizability. Many of these platforms, such as Moodle, Sakai, Docebo, Atutor, Ilias, LRN, etc., are increasingly used in schools and universities as a powerful support and improvement for teaching activities [1].

According to [2], using Moodle platform improved professor-student communication and increased students' satisfaction with their courses.

2 Methodology

Theoretical and empirical methods of research are applied for identifying the main regularities and productive achievements which can be further implemented into teaching practice. The following types of students' activities are used: observation, experiential learning, pedagogical documents analysis, studying the products of students' educational and creative activity, on-line surveys.

3 Results and discussion

3.1 On-line learning systems: advantages and disadvantages

Nowadays, there are a lot of disputes on the issue of on-line learning introduced to technical students (power engineering, space device engineering, electronic engineering, etc.). On the one hand, people supporting cutting-edge technologies distinguish their advantages which are mainly concerning their unlimited options and possibilities. On the other hand, there are quite a lot of people who single out more negative aspects in organizing self-study work using on-line learning, in particular in Moodle system. Let us carry out the analysis.

The main advantages of e-learning are the following. Firstly, there is no limitation in time and location. It implies that a student can study at any suitable time and place. Secondly, a student has the opportunity to ask the teacher questions in e-learning environment, particularly online. More than 30% of students expect their teacher's assistance and support. Moodle as an element of university e-learning environment allows the teacher to provide informational and conceptual support of the module. Thirdly, passing various tests and completing tasks in e-learning environment, students have the opportunity to obtain the results of their training, analyze their gaps in tasks as well as to study again the issues difficult for understanding. Fourthly, it is the way of realizing teacher's creative potential: Moodle allows including interactive and active teaching methods in the educational process (almost 40% of students expect an interesting way of presenting material): video lectures, creative tasks, tests, etc.

To the disadvantages we can refer the following: the necessity of available devices with the Internet connection as well as the situation when one cannot be sure that the task was carried out by the student individually without anybody's help. The obtained on-line experience is purely theoretical by nature. Besides, the development of students' communication skills can become complicated as the teacher does not always apply the active and interactive technologies.

During the educational process, the undergraduate's objective is to obtain the skills that would meet their future employer's requirements to the full extent. Nowadays, e-learning is becoming more and more popular among different companies. In terms of access, cost, function, online learning can be tailored in compliance with business needs. About 80% of employers often use technologies for training, retraining and professional development of their employees. New and experienced workers alike now had the opportunity to improve upon their industry knowledge base and expand their skill sets [3].

E-learning is beneficial to employers in many ways. Companies save between 50% and 70% of a budget when they replace an instructor-based training with e-learning. Training with e-learning refers courses delivered into shorter sessions and spread out over different days. It improves productivity, as employees no longer need to travel or get in rush-hour traffic to get to a class [4].

Empirical studies show that e-learning technology has positive effects on learning effectiveness and job performance as well. Furthermore, e-learning technology plays an important role in facilitating learning content and interacting with learners in organizational learning [5].

An increasing number of companies is relying on learning technology to boost their business forward. Towards Maturity Benchmark Survey shows that 72% of the 600 companies surveyed, claim that learning technologies helped their businesses adopt more quickly to change [4].

3.2 Description of the using experience

The increase in applying active and interactive teaching methods in e-learning environment has been already described above. Now we should consider the application of "Mind Maps" for organizing students' self-study work.

Mind Mapping is a method of capturing information created by Tony Buzan. It is believed that Mind Map can be applied to every aspect of life where improved learning and clearer thinking will enhance human performance.

For the purpose of designing tasks applying the method of Mind Maps, such Moodle tools as Seminar or Forum can be used. Applying these tools, the work performed by a student is evaluated by their fellow students and is further discussed.

Mind Maps can be designed using special software (fig. 1) as well as drawn by hand and uploaded to the system as a scan copy or photo (fig. 2).

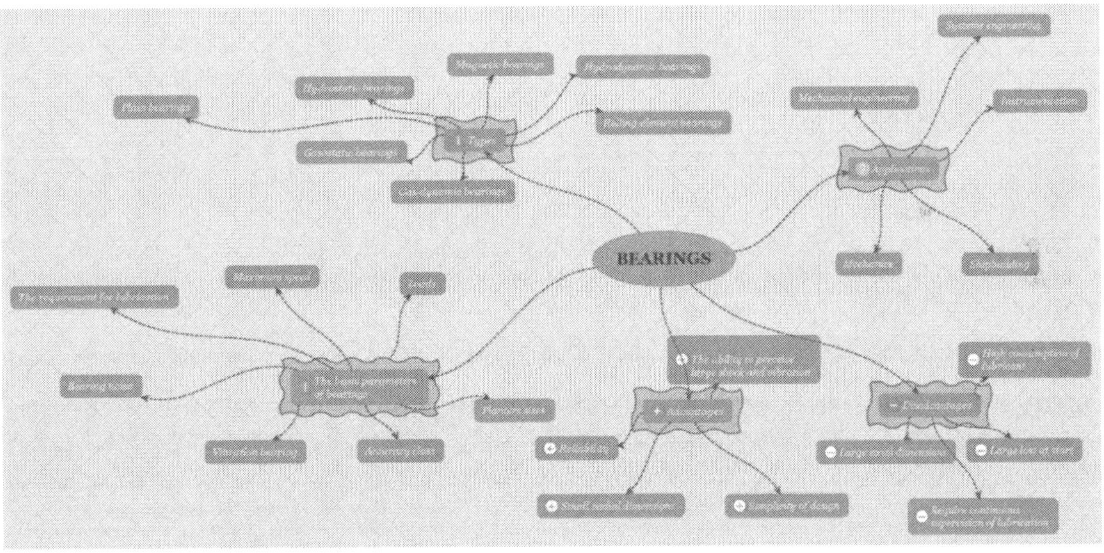

Figure 1. Mind map made by special program.

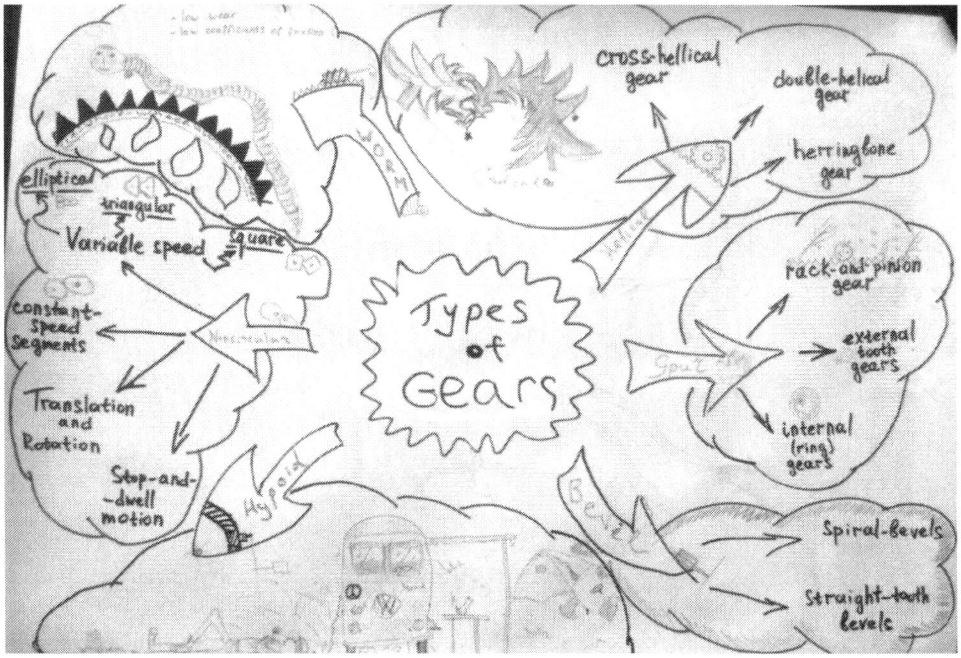

Figure 2. Hand-made Mind map.

The tool Forum can be used for organizing a group discussion. For example, in the course "Professional training in the English language" (module "Immediate problems of instrumentation engineering") the students were presented with 10 innovative solutions which, according to scientists, can improve the world. The students were allocated the task of choosing the three most important solutions from every student's point of view, discuss them with their fellow students and make a choice of the three most important solutions in the opinion of the whole group. It should be noted that, at the first glance, the easy task turned out to be quite difficult for accomplishing among Master's Degree students. Moreover, at the same time the similar task was allocated to the students during "face to face" classes (with the teacher present in the class). For an hour and a half this group could not arrive at a common view. In e-learning environment, the group of 12 students could not reach common ground as well. However, such a result of solving tasks is not a problem of working in e-learning environment and on-line communication as it identifies the necessity of developing students' communication skills.

A suggestion about the development of a practice-based approach concerns the introduction of electronic modules of the syllabus into technical programs of Bachelor, Master and Specialist's degrees. For example, in physics –solutions of equations, mathematic and physical modeling of processes; electronics – calculation of electric circuits, signal processing; in the design of devices and systems – 3D modeling of the aircraft structure, calculation of printed circuits, etc.

3.3 The results of the survey conducted among TPU students and teaching staff

In May-June 2015 TPU administration conducted a sociological survey. The purpose of the survey was to reveal the views of teachers about the use of TPU e-learning courses in the educational process. Ninety teachers of TPU were involved in the survey.

The results of the study showed that the attitude of respondents towards the use of e-learning courses in the educational process was generally positive. E-learning courses involve students in learning process and have a positive effect on their academic performance. "Test", "Task", "Glossary" and "Forum" were characterized as the most frequently used tools [7].

E-learning courses make the most sense being introduced in the curricula of Bachelor's and Master's degree programs (95,6%), facilitate effective management of students' self-study work

90,0%, additionally involve students into educational process 86,7% and have positive impact on students' performance 81,1%.

The teachers believe that the students are provided with a constant access to the educational materials (94,4%) and the possibility of obtaining additional bonuses (75,6 %); while performing some types of tasks, they do not treat it negatively (71,1%) and understand the necessity of using e-learning courses in the educational process. According to the obtained data, the most popular tools are "Test" 96,7%, "Task" 83,3%, "Glossary" 83,3% and "Forum" 72,2%.

More than one thirds of the respondents (37,8%) believe that such type of students' interaction as mutual assessment of their works does not bring them any discomfort, while 27,8% treat such experience quite negatively. The rest part of the respondents (34,4%) had some difficulty evaluating the process and expressing their opinion on this issue.

At the same time the survey of 268 TPU students was carried out. [8]

According to the students interviewed, the teachers interacted with them often within e-learning courses commenting on tasks that were performed in the classroom as well as assessing works submitted through LMS Moodle.

According to the students, the most useful features of e-learning is a constant access to learning materials/assignments, participation in online storage, assignments and the opportunity of asking their teacher questions at any time.

The majority of respondents indicate the potential of e-learning course directly for instruction and not for communication and interaction with their fellow students.

According to most students interviewed, e-learning course promotes preparation for the exam/test.

4 Conclusion

In the opinion of the majority of the teachers and students interviewed, the use of e-learning is advisable and will be designed both to boost learning motivation and student's performance as well as to facilitate more effective management of student's self-study work.

The objectives of active training in e-learning environment will not be achieved if joint activities and outcomes are not pre-planned.

It should be mentioned that along with knowledge obtained on professional topics the expertise and communication skills are developed as well.

Acknowledgement

This work is funded within the framework of realization of Strategic Programme on National Research Tomsk Polytechnic University Competitiveness Enhancement in the Group of Top Level World Research and Academic Institutions.

References

1. D. Benta, G. Bologa, I. Dzitac, PCS **31**, 1170 (2014)
2. Valentina Caputi, Antonio Garrido, JNCA **53,** 115 (2015)
4. 20 facts about e-learning. http://www.onlinecultus.com/20-facts-about-e-learning/ (2016)
5. Ecevit Zümrüt Sati, Dursun Faruk, DUJofSS **42**, 349 (2014)
6. Nina Bencheva, NTTRU – **49 3.2**, 63 (2010)
7. V. Ivanova, K. Mertins, EE **16**, 98 (2014)
8. *Mnenie prepodavatelei TPU ob ispol'zovanii ehlektronnyh kursov v uchebnom processe. Informacionno-analiticheskaya zapiska po rezul'tatam sociologicheskogo issledovaniya* (TPU 2015)
9. *Mnenie studentov TPU ochnoj formy obucheniya ob ispol'zovanii ehlektronnyh kursov v uchebnom processe. Informacionno-analiticheskaya zapiska po rezul'tatam sociologicheskogo issledovaniya* (TPU 2015)

MATEC Web of Conferences **48**, 06006 (2016)
DOI: 10.1051/matecconf/20164806006

Problem-based teaching in engineering education

Marina Kuimova[1,a], Douglas Burleigh[2] and Anastasiya Trofimova[1]

[1]*National Research Tomsk Polytechnic University, 634050 Tomsk, Russia*
[2]*La Jolla Cove Consulting, 92117 San Diego CA, USA*

Abstract. The aerospace industry requires highly educated, motivated engineers capable of working with increasingly complex processes and equipment, rapidly-changing requirements, and the need to constantly improve production efficiency. Colleges and universities throughout the world strive to provide training to young professionals for jobs in high-tech industries and to provide not only core technical knowledge, but also training on how to approach problems creatively and to generate novel solutions to problems. Problem-based learning (PBL) contributes to solving this problem. This paper reviews the literature on PBL, studies its benefits and drawbacks, and presents the positive results achieved by its implementation in the training of engineering students at Tomsk Polytechnic University.

1 Introduction

The traditional educational paradigm no longer meets the requirements of modern industry. The main reasons for this are the acceleration of social development, the changing situation in the labor market, and a decrease in formalization. Employers now expect more from job candidates to than just specialized education. Prospective employees have to be technically competent, resourceful, and independent to be competitive in the labor market. To satisfy the needs of business and industry, engineering universities must introduce innovative teaching methods and approaches in their curricula, and PBL (problem-based learning) is one of them.

Engineering universities strive to create a learning process that promotes self-education and self-development, and develops professional competence and creative skills. Creativity is required when students must face problems they have not learned in their education and need to find a way to solve these problems. PBL is a creative process that helps to develop skills for solving non-standard educational and scientific tasks. In the literature, PBL is described differently. We support the definition given by D. Boud & G. Feletti: "Problem-based learning is an approach to structuring the curriculum which involves confronting students with problems from practice which provide a stimulus for learning" [1]. In PBL, a teacher does not provide students with information on a particular issue, but organizes learning situations in which students solve problems and obtain new knowledge. It is a motivating and challenging approach, which has both followers and opponents. Supporters of PBL state that it enhances learning outcomes and:

- Allows learning by other than formal means;
- Teaches students to identify and search for needed knowledge,

[a]Corresponding author : mkuimova@mail.ru

- Provides meaningful context for knowledge;
- Promotes the abilities and skills to apply knowledge to new or unfamiliar situations;
- Encourages students to think deeply and independently, rather than just memorize facts;
- Develops problem-solving skills as well as self-directed learning skills;
- Improves interpersonal skills;
- Promotes creativity and independence;
- Promotes life-long learning [2-6].

Others believe that PBL may be less effective than traditional methods because it:

- Develops problem-solving skills at the expense of acquiring standard knowledge;
- Is inappropriate for beginners, as it requires a certain level of background knowledge;
- Requires time to find a solution to a problem;
- Does not provide a hierarchical list of topics;
- Makes monitoring and assessment difficult [7-10].

PBL empowers learners to conduct research, integrate theory with practice, and take responsibility for their own learning [11]. PBL is applied in a variety of educational institutions (medicine, architecture, engineering, business administration, law, psychology, nursing, etc.) and is integrated into a wide range of disciplines. In foreign language education, PBL helps learners acquire linguistic competence, develop abstract and situation-specific thinking, provokes reasoning, and trains students to solve unconventional tasks [12]. The purpose of this study is to examine the benefits of the implementation of PBL in teaching English as a foreign language in a university.

2 Discussion and research

The primary goal of PBL is to enhance learning by requiring students to identify and solve problems of a particular discipline. Problems are used as tools to acquire required knowledge. Learning occurs in small student groups. [13]. As a pedagogical tool, PBL helps solve a variety of educational tasks. It:

- Provides motivation to learn;
- Increases cognitive interest in educational and professional issues;
- Strengthens the retention of learned material;
- Promotes the independence of learners;
- Develops research skills;
- Develops creative abilities;
- Develops communicative competence [14-21].

PBL may be applied to a real problem (theoretical or practical issue) in a particular discipline. Students are introduced to the problem before they have learned the required knowledge. They work together to identify the learning requirements. It stimulates knowledge building, study and learning how reach a solution to a problem. Problems and learning situations must be carefully sequenced to ensure that students are taken through the syllabus. When selecting problems and learning situations, it is important to make sure that they are:

- Consistent with the syllabus;
- Clearly defined;
- Challenging, but possible to solve;
- Appropriate to the students' level of knowledge, intellectual abilities and interests.

A problem should be meaningful, and it should promote ideas and knowledge sharing. A problem should apply to science, a profession, or the learning process. It should also motivate students to find a solution to the problem, to perform research, and to recognize the possibility of ambiguous solutions. Students should also learn that there is rarely a unique approach to an investigation; there may be many.

The creation of a problem includes the following stages:

- Selection of educational material;
- Selection of questions and conditions that may evoke contradiction;

• Discussion of possible approaches to solve the contradiction.

In PBL, small student groups work on problem-based tasks. A teacher generally facilitates the learning process by guiding students, keeping them focused on the task, and stimulating interest, thought and discussion. The teacher also provides the educational materials, organizes the learning process, creates a stimulating learning environment, promotes the engagement of the whole group in discussions, and challenges the students to think creatively. In addition, the teacher monitors and assesses progress, and comments and summarizes progress. However, the teacher does not "teach" in the traditional way. The problem-solving competencies of the students are assessed with such techniques as case-study, self- and peer assessment, performance-based assessment, and the final report.

In general, the PBL process includes the following stages:

- Reading the task and defining the problem;
- Brainstorming possible solutions;
- Arrange explanations and define the learning objectives needed to test/prove the validation of assumptions and hypotheses;
- Study available information resources;
- Share the results of the study with the group.

To evaluate the effectiveness of PBL, we created two groups of students, each consisting of sixteen fourth-year students studying English for Electrical Engineering at National Research Tomsk Polytechnic University (TPU). One group was a "control" and the other was the experimental group. For the whole semester, the students of the control group were taught using traditional methods and the students of the experimental group were taught using PBL. At the end of the semester, all of the students were asked about their attitudes on their experience with the program, and they were asked to indicate on a scale from 1-6 (where 1 is the highest mark) their preference for the form of work in their classes:

1. Individual (independent) work;
2. Teacher-centric activities;
3. Pair work;
4. Group work;
5. Research work;
6. Discussions.

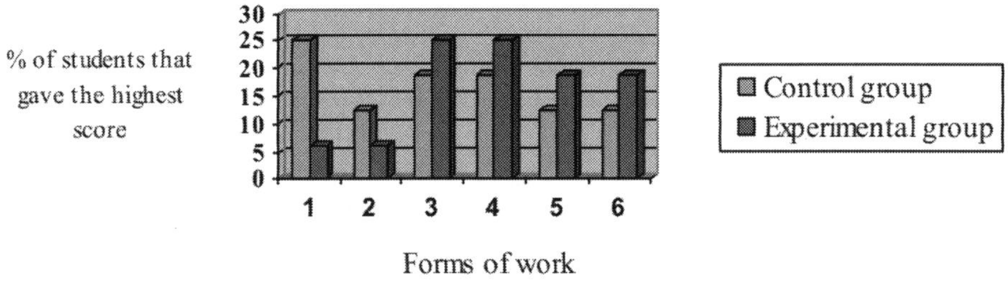

Figure 1. Student preferences for the forms of work in classes.

An analysis of the study on the forms of work used in classes has shown that students in the experimental group prefer innovative methods (pair work, group work, research work, discussions).

At the beginning, middle and end of the semester, the students of both groups performed assessment tasks. The groups were given the same tests. The results of the tests are presented on the histogram in Figure 2. The Y-axis shows the average grade received by the group (10 was the maximum score), and the X-axis shows which test.

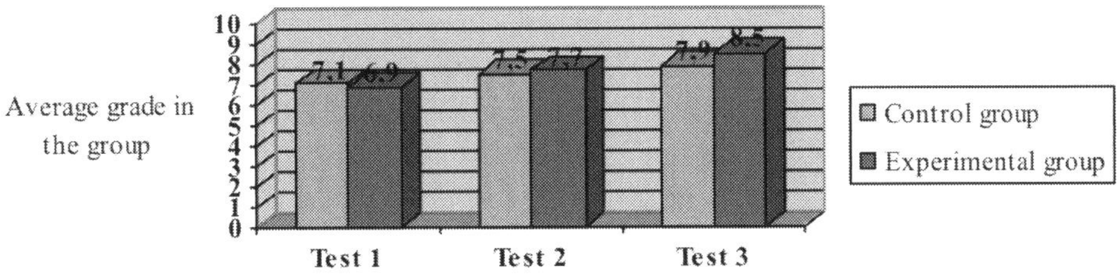

Figure 2. Assessment of students' achievements.

The histogram shows that at the start of the semester the average grade of the control group was 7.1, which is slightly higher than the 6.9 score of the experimental group. However, at the end of the semester the experimental group showed better achievement, as the average grade or the control group was 7.9, while the average grade of the experimental group was 8.5.

The results showed that PBL contributes to more favorable conditions for learning.

3 Conclusion

Higher education is a driving force for innovation, economics, industry and business. It is a factor in a country's competitiveness in the global market. Innovative teaching methods and approaches at universities are aimed at developing students' potentials and to prepare them for lifelong learning. In this paper we have evaluated PBL in the context of its educational opportunities in teaching English to engineering students. The research proves that PBL motivates learning, enhances learning outcomes, improves research, and should increase career benefits.

References

1. D. Boud, G. Feletti, *The challenge of problem-based learning (2nd ed.)*, 334 (1997)
2. R.E. Mayer, Educ Psychol Rev. **8(4)**, 357 (1996)
3. E.De Graaf, A. Kolmos. IJEE **19(5)**, 657 (2003)
4. Ch.Chin1, Li-Gek Chia, Sci Educ. **88(5)**, 707 (2004)
5. A. Masek, S. Yamin, Int. rev. soc. sci. humanit. **2(1)**, 215 (2011)
6. D.H. Jonassen, W. Hung, *Problem-Based Learning*. Encyclopedia of the Sciences of Learning, edited by N.M. Seel, 2687 (2012)
7. E.M. Lieux. *A skeptic's look at PBL. In The Power of Problem-Based Learning: A Practical 'How To' for Teaching Undergraduate Courses in Any Discipline, edited by B. Duch, S.E. Groh & D.E. Allen*, 223 (2001)
8. C. Angeli, Sci. Educ. Int. **13(3)**, 9 (2002)
9. K.D., Simons, J.D., Klein, T.R. Brush, JILR **15(3)**, 213 (2004)
10. P.A. Kirschner, J. Sweller, R.E. Clark, Educ. Psychol. **41(2)**, 75 (2006)
11. J.R. Savery, IJPBL **1(1),** 9 (2006)
12. L. Heine. *Problem Solving in a Foreign Language: A Study in Content and Language Integrated Learning*. Berlin: De Gruyter, (2010)
13. D. Gijbels, F. Dochy, P. van den Bossche, M. Segers, Rev. Educ. Res. **75(1)**, 27 (2005)
14. J.R. Mergendoller, N.L. Maxwell, Y. Bellisimo, J. Educ. Res. **93(6)**, 374 (2000)
15. C.E. Hmelo-Silver, Educ Psychol Reve **16(3)**, 235 (2004)
16. J.C. Perrenet, P.A.J. Bouhuijs, J.G.M.M. Smits, Teach High Educ. **5(3)**, 345 (2010)
17. A. Klegeris, H. Hurren, Adv Physiol Educ. **35(4)**, 408 (2011)
18. A. EL-Shaer, H. Gaber, JEP **5(14)**, 74 (2014)

19. M.V. Kuimova, D.A. Polyushko, (iJET). **10(1)**, 72 (2015).
20. A.V. Ruchina, M.V. Kuimova, D.A. Polyushko, A.E. Sentsov, Zhang Xue Jin, Procedia Soc Behav Sci. **215**, 98 (2015)
21. A. Klegeris, M. Bahniwal, H. Hurren, CBE Life Sci Educ. **12(1)**, 73 (2013)